H = φ

UNIVERSITY MATHEMATICAL SERIES

General Editor: G. T. Kneebone, *Bedford College, London*

MATHEMATICAL LOGIC
AND HILBERT'S ε-SYMBOL

MATHEMATICAL LOGIC
AND
HILBERT'S ε-SYMBOL

A. C. LEISENRING
Reed College, Portland, Oregon

MACDONALD TECHNICAL & SCIENTIFIC
LONDON

MACDONALD & CO. (PUBLISHERS) LTD
49/50 POLAND STREET, LONDON W.1

PRINTED AND BOUND IN ENGLAND BY
HAZELL WATSON AND VINEY LTD
AYLESBURY, BUCKS

PREFACE

This study of Hilbert's ε-symbol is based on a doctoral dissertation which was submitted to the University of London in June 1967. It contains a number of new results, in particular a strengthening of Hilbert's Second ε-Theorem, which should be of some interest to specialists in the field of mathematical logic. However, this book is not written for the expert alone. We have tried to make it as self-contained as possible so that most parts will be intelligible to anyone with a good mathematical background but with a limited knowledge of formal logic. Since the ε-symbol can be used to overcome many technical difficulties which arise in formalizing logical reasoning, the present book may be useful as supplementary reading material for an undergraduate course in logic.

In Chapter I we define the formal languages which are used throughout the book and analyze the semantics of such languages. An interesting result concerning finitary closure operations is then established. This result is used in Chapter II to prove the (semantic) completeness of certain formal systems which incorporate the ε-symbol. In Chapter II we also establish a number of derived rules of inference for these formal systems. Chapter III, which is more technical than Chapter II, contains proofs of Hilbert's ε-Theorems, Skolem's Theorem, and Herbrand's Theorem. If the reader is interested in only these results, he may omit all of Chapter I except for the definitions of formal languages and may skim through most of Chapter II.

Chapter IV deals with formal theories. We see how the ε-symbol and ε-Theorems may be used to prove the consistency of various mathematical theories. In particular, Hilbert's attempts to prove the consistency of arithmetic are described. We then discuss the role which the ε-symbol can play in formalizing set theory, and we give particular attention to the relationship between this symbol and the axiom of choice.

Chapter V may be regarded as an appendix to the book. In this chapter we give alternative proofs of some of the theorems in Chapter III. Our methods illustrate the close connection which exists between Hilbert's ε-Theorems and Gentzen's *Hauptsatz*.

The author would like to express his gratitude to G. T. Kneebone (Bedford College, London) under whose helpful supervision his thesis was written. He is also indebted to Marian Cowan for her excellent work in typing the manuscript, and most of all to his wife, Flora, for her help, patience, and encouragement.

A. C. Leisenring

Reed College

Portland, Oregon

CONTENTS

Chapter III   The ε-Theorems

Chapter IV   Formal Theories

INTRODUCTION

1 Objectives

The purpose of this book is to examine the nature of Hilbert's ε-symbol and to demonstrate the useful role which this logical symbol can play both in proving many of the classical theorems of mathematical logic and also in simplifying the formulation of logical systems and mathematical theories.

The ε-symbol is a logical constant which can be used in the formal languages of mathematical logic to form certain expressions known as ε-terms. Thus, if A is a formula of some formal language $\mathscr{L}$ and x is a variable of $\mathscr{L}$, then the expression $\varepsilon x A$ is a well-formed term of the language. Intuitively, the ε-term $\varepsilon x A$ says 'an x such that if anything has the property A, then x has that property'. For example, suppose we think of the variables of the language as ranging over the set of human beings and we think of A as being the statement 'x is an honest politician'. Then $\varepsilon x A$ designates some politician whose honesty is beyond reproach, assuming of course that such a politician exists. On the other hand, if there are no honest politicians, then $\varepsilon x A$ must denote someone, but we have no way of knowing who that person is. Similarly, even if there are honest politicians, we have no way of knowing which one of them $\varepsilon x A$ designates.

Since the ε-symbol, or ε-operator as it is sometimes called, selects an arbitrary member from a set of objects having some given property, this symbol is often referred to as a 'logical choice function'. It is not surprising then that an investigation of the ε-symbol also sheds some light on the nature of the axiom of choice. One of the main theorems of this book, Hilbert's Second ε-Theorem, provides a formal justification of the use of the ε-symbol in logical systems by showing that this symbol can be eliminated from proofs of formulae which do not themselves contain the symbol. What this theorem says intuitively is that the act of making arbitrary choices is a legitimate logical procedure. However, it has been shown by Cohen [1966] that an application of the axiom of choice in set theory cannot in general be eliminated. It follows then that the real power of the axiom of choice lies not in the fact that it allows one to make arbitrary selections but rather in the fact that it asserts the existence of a set consisting of the selected entities.

Before saying anything more about the nature of the ε-symbol and the role it plays, we should first give a brief explanation of the basic objectives and concepts of mathematical logic.

2   Formal languages

The objects of study in mathematical logic are formal or symbolic languages, and the primary aim is to provide a formalization of logical reasoning in terms of such languages. A *formal language* $\mathscr{L}$ consists of a set of symbols together with certain prescribed rules for building well-formed (or grammatical) expressions from these symbols. The well-formed expressions of $\mathscr{L}$ are usually divided into two categories, *formulae* and *terms*. The formulae correspond to the sentences of ordinary languages and the terms correspond to the nouns, pronouns, and noun clauses. The symbols of the language include certain function and predicate symbols (the *vocabulary* of the language), variables, and logical constants. To the logician the most important symbols are the logical constants. These usually consist of certain *propositional connectives*, such as $\lnot$ ('not'), $\lor$ ('or'), $\land$ ('and'), $\to$ ('implies'), and $\leftrightarrow$ ('if and only if'), and one or both of the *quantifiers*, $\forall$ ('for all') and $\exists$ ('there exists'). The formal languages dealt with in this book also contain the logical constant ε (Hilbert's ε-symbol).

To avoid any confusion between the formal language being studied and the (informal) language which is used in carrying out this study, one refers to the former as the *object language* and to the latter as the *metalanguage*. Similarly, the mathematical theory which is used in reasoning about the object language is called the *metatheory*. Ideally, the metatheory should be as weak as possible. For example, a proof of the consistency of formal arithmetic would carry little weight if one were to use a strong metatheory which included all of arithmetic and set theory. Therefore, throughout this book whenever possible we shall use a weak 'constructive' metatheory, even though certain theorems can be proved more easily using 'non-constructive' techniques. (A non-constructive argument is one in which the existence of something is proved by deducing a contradiction from the assumption that no such thing exists.)

Having specified a formal language $\mathscr{L}$, one can then formalize logical reasoning in either of two ways, (i) in terms of 'models', or (ii) in terms of 'formal systems'.

2.1   The semantic consequence relation ⊨

Without going into too much detail we can say that a *model* is an abstract mathematical structure which provides an interpretation of the symbols of $\mathscr{L}$ in such a way that every formula of $\mathscr{L}$ becomes either a true or a false statement about this structure. The logical constants of $\mathscr{L}$ are always given their natural interpretation, but the interpretations given to the function and predicate symbols may differ for different models. Thus a formula A may be true in one model and false in another. A formula A is said to be a semantic consequence of a set of formulae X, denoted by $X \vDash A$, if A is true in every model in which all the members of X are true. For example, suppose that $\mathscr{L}$

is the language of elementary group theory and X is the set of formal axioms of this theory. Then to say that A is a semantic consequence of X amounts to saying that A is a statement which holds true for any group G. Obviously if $\emptyset \vDash A$, where $\emptyset$ is the empty set, then A is true in all models, and we then say that A is *valid* (or *logically true*).

2.2 The deducibility relation ⊢

The semantic consequence relation, $\vDash$, provides a useful, though rather abstract, formalization of logical consequence and logical truth. One objection to this approach is that one needs a strong metatheory which permits non-constructive arguments about infinite sets. For many reasons it is often preferable to regard the formulae of a language $\mathscr{L}$ as concrete objects, i.e., uninterpreted sequences of symbols, and to confine one's metatheory to constructive, combinatorial arguments about these objects.

For this reason one often formalizes logical reasoning in terms of 'deducibility' in a formal system $\mathscr{F}$, where a formal system usually consists of certain axioms and rules of inference for the language $\mathscr{L}$. The notation $X \vdash_{\mathscr{F}} A$ is used to denote that A is deducible from X in $\mathscr{F}$. A formal system $\mathscr{F}$ is said to be *sound* and (*semantically*) *complete* if the deducibility relation ⊢ for $\mathscr{F}$ coincides with the semantic consequence relation $\vDash$. In other words, by proving that $\mathscr{F}$ is both sound and complete one justifies the axioms and rules of inference of $\mathscr{F}$.

The standard formal system used in mathematical logic is the (first-order) *predicate calculus*, and proofs of the soundness and completeness of this system are included in most textbooks of mathematical logic. The primary formal system which is dealt with in this book is called the *ε-calculus*. This system is essentially obtained from the predicate calculus by adjoining the ε-symbol as a new logical constant and by introducing some additional axioms for dealing with this symbol. It is our contention that by enlarging the predicate calculus in this way one obtains a much neater and simpler formalization of logical reasoning.

We have now introduced enough of the basic concepts to give a precise statement of Hilbert's Second ε-Theorem. Suppose A is a formula of some language $\mathscr{L}$, X is a set of formulae of $\mathscr{L}$, and the ε-symbol does not occur in A or in any member of X. The Second ε-Theorem states that if A is deducible from X in the ε-calculus, then A is deducible from X in the predicate calculus. A more succinct statement of the theorem would be that the ε-calculus is an 'inessential extension' of the predicate calculus.

2.3 Formal theories and the formalist programme

Strictly speaking, mathematical logic is a branch of applied mathematics. However, unlike other branches of applied mathematics which are used to solve problems in the natural and social sciences, mathematical logic is used to prove results about mathematical theories. When mathematical logic is used in this way, the subject is often referred to as *metamathematics*.

A formalization $\mathcal{T}$ of a particular mathematical theory, e.g. arithmetic or set theory, consists of a formal language $\mathcal{L}$ which is adequate for expressing the concepts of that theory, a suitable formal system $\mathcal{F}$, such as the predicate calculus, and a specified set X of formulae of $\mathcal{L}$ which serve as the (non-logical) axioms of the theory. A formula A of the language $\mathcal{L}$ is said to be a *theorem* of the formal theory $\mathcal{T}$ if A is deducible from X in $\mathcal{F}$. The theory $\mathcal{T}$ is said to be *inconsistent* if a contradictory formula, i.e., one of the form $A \wedge \neg A$, is a theorem of $\mathcal{T}$; otherwise $\mathcal{T}$ is said to be *consistent*.

By reducing a mathematical theory to a formal theory one can prove various results about that theory in a completely constructive way, since a formal theory is nothing more than a meaningless array of symbols, together with certain prescribed rules for manipulating these symbols.

This study of mathematical theories within a constructive metatheory was first developed by the *formalists* under the leadership of David Hilbert. The discovery of certain paradoxes in set theory around the year 1900 had aroused grave doubts about the legitimacy of the non-constructive techniques which mathematicians often used in dealing with infinite sets. The formalists were convinced that these techniques were justifiable, and they hoped to find such a justification by proving constructively that the basic mathematical theories, such as arithmetic and analysis, were consistent. Unfortunately for them, it was shown by Gödel [1931] that even for such a simple theory as arithmetic no such consistency proof can ever be found.

Despite the fact that their primary goal proved to be unattainable, the formalists made outstanding contributions to the theory of mathematical logic, and paved the way for many important later discoveries.

3 The history of the ε-symbol

The ε-symbol was introduced by Hilbert and his collaborators in order to provide explicit definitions of the quantifiers ∀ and ∃. These definitions are expressed by the formulae

(1) $$\exists x A \leftrightarrow A(\varepsilon x A)$$

and

(2) $$\forall x A \leftrightarrow A(\varepsilon x \neg A),$$

where $\neg$ is the symbol for 'not'. Hilbert was convinced that by using the ε-symbol rather than the quantifiers in formalizing arithmetic and analysis, one could establish the consistency of these two theories. The first published work in which the ε-symbol is used is Ackermann's doctoral dissertation [1924], written under Hilbert, in which an attempt is made to prove the consistency of analysis. However, in the previous year an article by Hilbert [1923] appeared in which a similar symbol, the τ-symbol, is used. A comprehensive account of the results which the formalists proved using the ε-symbol is given by Hilbert and Bernays [1939].

The ε-calculus used by Hilbert and Bernays is essentially the formal system which is obtained from the predicate calculus by adjoining the ε-symbol as an additional logical constant and by taking all formulae of the form

$$(3) \qquad\qquad A(t) \rightarrow A(\varepsilon x A)$$

as additional axioms, where t is any term. It can then be shown that (1) and (2) follow from (3) by virtue of the axioms and rules of inference of the predicate calculus. The two main results which Hilbert and Bernays prove concerning the ε-symbol are known as the First and Second ε-Theorems. The first of these is concerned with the eliminability of the quantifiers from the predicate calculus, and as we have already seen, the second is concerned with the eliminability of the ε-symbol from the ε-calculus.

For the most part, Hilbert and Bernays use their ε-calculus only in a subsidiary role to prove that certain deductions in the predicate calculus can be rewritten in a simpler form. However, an ε-calculus can be used advantageously as a formal system in its own right. Ackermann [1937–8] and Bourbaki [1954] present interesting formalizations of set theory which are based on an ε-calculus. The axioms of their ε-calculus include all formulae of the form

$$(4) \qquad\qquad \forall x(A \leftrightarrow B) \rightarrow \varepsilon x A = \varepsilon x B$$

in addition to formulae of the form (3) above. When this formal system is used in formalizing set theory, there is usually no need to adopt the axiom of choice since this axiom is deducible from an axiom of replacement using axiom (3) above. The conditions under which the axiom of choice is deducible will be discussed in Chapter IV (see page 106).

Naturally, if the ε-calculus is used in this way as a formal system in its own right, one would like to know whether the system is sound and complete. The answer to this question depends on the semantic interpretation which is given to the ε-symbol. Asser [1957] in his Berlin *Habilitationsschrift* interprets the ε-symbol as a 'choice function' and proves the soundness and completeness of various forms of the ε-calculus. More recently, Hermes [1965] proves the same result for his *Termlogik mit Auswahloperator*. In our proofs of the soundness and completeness of the ε-calculus we shall also interpret the ε-symbol as a choice function.

4 The indeterminacy of the ε-symbol

One of the most intriguing and useful features of the ε-symbol is its indeterminacy. Carnap [1961], pages 162–163, describes this feature as follows:

'The symbol "ε" was intentionally introduced by Hilbert as an indeterminate constant. Its meaning is specified by the axioms only to the extent that any non-empty set has exactly one representative and this representative is an element of the set. If the set has more than one element, then nothing is said, either officially or unofficially, as to which of the elements is the representative. Thus, for example, $\varepsilon x(x = 1 \lor x = 2 \lor x = 3)$ must be either 1, or 2, or 3; but there is no way of finding out which it is.'

We might add to Carnap's remarks that the situation is even more mystifying if there is no x such that A is true. For example, suppose A is the formula $x \neq x$; then the term $\varepsilon x A$ must denote some object, but we have no way of knowing what that object is. A term such as this is called a *null term*.

We shall see that the usefulness of the ε-symbol is due essentially to its indeterminate nature. For example, because of its indeterminacy the ε-symbol provides useful derived rules of inference for the elimination and introduction of quantifiers (see page 48) and enables one to give explicit definitions of indefinite concepts, such as the concept of cardinal number in set theory (see page 104). However, this indeterminacy also accounts for the suspicion with which the ε-symbol is often regarded. We hope that this book will dispel some of that suspicion.

5   Possible ways of defining formal languages

We shall conclude these introductory remarks with an explanation of the motivation behind the particular way in which we have defined our formal languages.

In setting up a formal language $\mathscr{L}$ it is not necessary to take all the logical constants as symbols of the language, since some may be defined in terms of others. For example the symbol $\leftrightarrow$ may be defined in terms of $\rightarrow$ and $\land$ by regarding an expression of the form $A \leftrightarrow B$ as an abbreviation for the expression $(A \rightarrow B) \land (B \rightarrow A)$. The actual logical symbols of the language are called *primitive* symbols and those symbols which are defined in terms of the primitive symbols are called *defined* symbols.

Our formal languages have the following logical (primitive) symbols: $\int$ (for some false proposition), $\neg$, $\rightarrow$, $\land$, $\lor$, $\exists$, $\forall$, $=$, and ε. The only defined symbol is $\leftrightarrow$. It is hoped that by defining our languages with such a rich array of symbols our results are more general than they would be if we were more selective in our choice of logical primitives. In fact, all our main results would still hold if our languages had only a certain subset of these symbols from which the remaining symbols could be defined.

In particular, the symbol $\int$ could be dispensed with by choosing some fixed formula A_0 of $\mathscr{L}$ and defining $\int$ as $A_0 \land \neg A_0$. Conversely, the negation symbol $\neg$ could be defined by letting $\neg A$ be $A \rightarrow \int$. For our purposes it is advisable that at least one of the symbols f and $\neg$ be taken as a primitive. With regard to the binary connectives $\rightarrow$, $\wedge$, and $\vee$, it is possible to define any two of these in terms of the third and the negation symbol. Since we shall be using an elegant unifying notation, due to Smullyan [1965], whereby certain formulae are classified as 'conjunctive' or 'disjunctive' formulae, it will be apparent that all our results would hold if the languages under consideration contained only one or two of these three connectives. Similarly, because of Smullyan's notation, our results would hold if only one of the quantifiers were taken as a primitive and the other were defined in terms of it.

We have seen (page 4) that both the quantifiers can be defined in terms of the ε-symbol. However, since we shall be dealing with the predicate calculus, where the ε-symbol is unavailable, it is necessary to take at least one of the quantifiers as a primitive. Furthermore, the definition of the quantifiers in terms of the ε-symbol leads to certain technical difficulties involving the relabelling of bound variables. Although these difficulties can be overcome in various ways, we shall not deal with them in this book. Methods for defining the quantifiers in terms of the ε-symbol appear elsewhere in the literature. See for example, Bourbaki [1954], Hermes [1965], and a paper by the author [1968] in which the main results of Chapter I of this book are proved for languages in which the only logical primitives are f, $\rightarrow$, and ε.

In stating the rules of formation of our languages we do not exclude the possibility of 'vacuous bondage' or 'collisions of bound variables'. Thus an expression of the form $\exists xA$, $\forall xA$, or εxA may be well-formed even if the variable x has no free occurrence in A (vacuous bondage) or even if x already has a bound occurrence in A (collision of bound variables). Nowadays, many logicians allow these two situations because in so doing, various technical difficulties can be avoided, and no new complications arise. For example, we can replace one ε-term by another in a formula A without first relabelling the bound variables in A to avoid any collisions which might result.

With regard to the actual symbols which we use for individual objects it has proved convenient to have two sorts available: (i) the 'variables' $x_1, x_2, \ldots$ which are denoted syntactically[1] by the letters u, v, w, x, y, and z, and (ii) the 'individual symbols' $a_1, a_2, \ldots$ denoted by a, b, and c. The variables may be bound by the operators $\exists$, $\forall$, and ε, whereas the individual symbols are never bound. The terms and formulae of a language are defined as those well-formed expressions in which no variable has a free (unbound) occurrence. So far, our formulation does not differ from that of Hilbert and Bernays and others who use different symbols for 'free variables' and 'bound variables'. Our approach differs from the Hilbert-type formulation in that we follow the more modern attitude of interpreting the symbols, $a_1, a_2, \ldots$, as 'arbitrary

---

[1] In other words, the letters u, v, $\ldots$ are used in the metalanguage as names for unspecified variables.

constants' rather than as 'free variables'. This means that within our formal system there is no general rule of substitution for individual symbols. There are the derived rules that from $X \vdash A(a)$ one can infer $X \vdash A(t)$ and $X \vdash \forall x A(x)$ provided that a does not appear in any member of X. However, if a does appear in the assumption set X, then the interpretation of a is temporarily 'fixed' and it cannot be treated as a free variable is treated in a Hilbert-type system. One advantage of this approach is that the statement of the deduction theorem does not involve any complicated restrictions concerning the use of the individual symbols, since these restrictions are already built into the system.

It should be pointed out that instead of specifying a list of individual symbols in our definition of a language, we could require that the vocabulary must contain infinitely many 0-place function symbols. This method is used by Robinson [1963]. These function symbols, or constants, could then be used as the individual symbols. We have not taken this approach since it seems more natural to confine the 0-place function symbols of the vocabulary to those particular constants, such as the symbol for zero in arithmetic, which have a fixed intended interpretation. Within a deduction, the individual symbols do not behave as constants since they are not the names of definite objects. Consequently, these symbols should be regarded more as logical symbols than as vocabulary symbols. In fact, we shall see that the individual symbols in a deduction can be thought of as abbreviations for ε-terms. The close connection between these symbols and ε-terms is brought out by the fact that in formal systems which incorporate the ε-symbol, the individual symbols can be dispensed with, since their role is assumed by the ε-terms of the language.

CHAPTER I

SYNTAX AND SEMANTICS

1 Introduction

In the study of ordinary languages, such as English, the word 'syntax' is used to refer to the grammatical construction of sentences. Similarly, we define the *syntax* of a formal language $\mathscr{L}$ to be the basic structure of $\mathscr{L}$ as laid down by its rules of formation. Thus a syntactical study of $\mathscr{L}$ officially ignores any intended interpretations of the symbols of $\mathscr{L}$. For example, the notion of formula is a syntactic concept, whereas the notions of validity and theorem are not. On the other hand, a *semantic* study of a formal language $\mathscr{L}$ deals with the possible interpretations which may be given to the symbols, terms, and formulae of $\mathscr{L}$. Thus the notions of model, validity, and satisfiability are semantic concepts.

This chapter deals with the syntactic and semantic properties of formal languages which incorporate Hilbert's ε-symbol. The main theorem of the chapter, Theorem I.11, establishes an abstract semantic property of such formal languages which is of sufficient generality for both the Compactness Theorem and the Completeness Theorem for the ε-calculus to follow from it as corollaries. The Compactness Theorem, which has many applications in other branches of mathematics, can be stated as follows:

For any set X of formulae of $\mathscr{L}$, if every finite subset of X has a model, then X has a model whose cardinality is less than or equal to the cardinality of the set of symbols of $\mathscr{L}$.

It is well known that Henkin's proof of the completeness of the predicate calculus (Henkin [1949]) can be used to prove the Compactness Theorem. However, since this theorem deals only with models of sets of formulae, it is natural to look for a proof which is purely model-theoretic—that is, one which does not depend on the Completeness Theorem or on the particular set of axioms and rules of inference which have been chosen to give the language a formal deductive structure. One advantage of our abstract approach is that the resulting proof of the Compactness Theorem is model-theoretic in this sense.

In order to carry out our semantic investigations it is necessary to use ordinary mathematical reasoning about sets. Consequently, throughout this chapter our metatheory is set theory with the axiom of choice. The set-theoretic symbols $\in$, $\emptyset$, $\subseteq$, $\cup$, $\cap$, $\backslash$, and $\{x: \ldots x \ldots\}$ are used in the usual way.

2   The formal language $\mathscr{L}(\mathscr{V})$

A *vocabulary* $\mathscr{V}$ is an ordered triple $\langle Fn, Pr, \rho \rangle$ where Fn and Pr are any two disjoint sets and ρ is a function from $Fn \cup Pr$ into the set of non-negative integers. The elements of Fn are called the *function symbols* of $\mathscr{V}$ and the elements of Pr are called the *predicate symbols* of $\mathscr{V}$. For any g in Fn or P in Pr, $\rho(g)$ is called the *order* of g and $\rho(P)$ is called the *order* of P. Any function symbol (predicate symbol) of order n is called an *n-place* function symbol (*n-place* predicate symbol). A 0-place function symbol is often called a *constant* and a 0-place predicate symbol is often called a *proposition*. The letters g and h, with or without subscripts, will be used as metalinguistic variables to denote arbitrary function symbols, and in particular, g^n will be used to denote an arbitrary n-place function symbol. The letter P, with or without subscripts, will be used to denote an arbitrary predicate symbol, and P^n to denote an arbitrary n-place predicate symbol. We place no restriction on the cardinality of the sets Fn and Pr. In particular, both may be empty.

Let $\mathscr{V}$ be the vocabulary $\langle Fn, Pr, \rho \rangle$ and let $\mathscr{V}'$ be the vocabulary $\langle Fn', Pr', \rho' \rangle$. We say that $\mathscr{V}'$ is an *extension* of $\mathscr{V}$ if $Fn \subseteq Fn'$, $Pr \subseteq Pr'$, and ρ and ρ' agree on $Fn \cup Pr$. Given a vocabulary $\mathscr{V}$, we often form an extension of that vocabulary by *adjoining* new function symbols or predicate symbols. For example, to formalize arithmetic we might use a vocabulary $\mathscr{V}$ consisting of the constants 0 and 1, and the 2-place function symbols $+$ and $\cdot$. The vocabulary $\mathscr{V}'$ obtained from $\mathscr{V}$ by adjoining the 2-place predicate symbol $<$ is then an extension of $\mathscr{V}$.

We shall now define the unique formal language which is determined by a given vocabulary $\mathscr{V}$. We normally denote this language by $\mathscr{L}(\mathscr{V})$. However, if it is irrelevant to our discussion what the vocabulary $\mathscr{V}$ is, we write $\mathscr{L}$ instead of $\mathscr{L}(\mathscr{V})$.

The *symbols* of $\mathscr{L}(\mathscr{V})$ are:

1. the function symbols and predicate symbols of $\mathscr{V}$;
2. the variables x_1, x_2, x_3, $\ldots$;
3. the individual symbols a_1, a_2, a_3, $\ldots$;
4. the logical constants f, $\neg$, $\rightarrow$, $\wedge$, $\vee$, $\exists$, $\forall$, ε, and $=$;
5. the separation symbols $($, $)$, and $,$.

We require that no symbol comprehended under any one of the clauses 1, 2, 3, 4, and 5 is comprehended under any other. The letters u, v, w, x, y, z, with or without subscripts, will be used to denote arbitrary variables, and the letters *a*, *b*, *c*, with or without subscripts, to denote arbitrary individual symbols. The set of variables will be denoted by *Vr* and the set of individual symbols by *Ind*.

An *expression of length n* is any string $s_1 \ldots s_n$ of symbols of $\mathscr{L}(\mathscr{V})$. We include the possibility that $n = 0$. In this case the expression is called the *empty expression*, and we denote this expression by *e*. Given any two expressions *A* and *B*, we can form a new expression *AB* by juxtaposing *A* and *B*. Thus if *A* is an expression of length *m* and *B* is an expression of length *n*, then the length of *AB* is $m + n$. We say that an expression *A occurs in* (or *appears in* or *is contained in*) an expression *B* if *B* is of the form $B_1 A B_2$ where B_1 and B_2 are any two expressions.

The *well-formed expressions* of $\mathscr{L}(\mathscr{V})$ fall into two disjoint categories, the *quasi-terms* and the *quasi-formulae*. These are defined by the following recursive *rules of formation*.

G1. Any variable or individual symbol is a quasi-term.

G2. If g^n is an *n*-place function symbol of $\mathscr{V}$ ($n \geqslant 0$), and $t_1, \ldots, t_n$ are quasi-terms, then $g^n t_1 \ldots t_n$ is a quasi-term.

G3. If P^n is an *n*-place predicate symbol of $\mathscr{V}$ ($n \geqslant 0$), and $t_1, \ldots, t_n$ are quasi-terms, then $P^n t_1 \ldots t_n$ is a quasi-formula.

G4. If *s* and *t* are quasi-terms, then $s = t$ is a quasi-formula.

G5. The symbol f is a quasi-formula.

G6. If *A* and *B* are quasi-formulae, then $\neg A$, $(A \to B)$, $(A \wedge B)$, and $(A \vee B)$ are quasi-formulae.

G7. If *A* is a quasi-formula, then for any variable *x*, $\exists x A$ and $\forall x A$ are quasi-formulae.

G8. If *A* is a quasi-formula, then for any variable *x*, $\varepsilon x A$ is a quasi-term.

G9. Only those expressions which are generated by G1–G8 are well-formed.

Any expression which is well-formed by virtue of G8 is called a *quasi ε-term* and any which is well-formed by virtue of G3, G4, or G5 is called an *atom*. A well-formed expression is said to be *ε-free* if the symbol ε does not occur in it, *identity-free* if the symbol = does not occur in it, and *elementary* if the symbols ∀, ∃, and ε do not occur in it.

Unless otherwise stated, the letters *A*, *B*, and *C*, with or without subscripts, will be used to denote arbitrary quasi-formulae, and the letters *s* and *t*, with or without subscripts, will be used to denote arbitrary quasi-terms. The following syntactic abbreviation is used throughout:

$$(A \leftrightarrow B) \quad \text{for} \quad ((A \to B) \wedge (B \to A)).$$

We shall adopt the following conventions for the omission of parentheses. First of all, we normally omit the outermost pair of parentheses. For example, we write

$$(A \wedge B) \to C$$

instead of
$$((A \wedge B) \to C).$$
Secondly, we adopt the rule of association to the right. This means that $A \vee B \vee C, A \wedge B \wedge C$, and $A \to B \to C$ should be read as $(A \vee (B \vee C))$, $(A \wedge (B \wedge C))$, and $(A \to (B \to C))$, respectively. Thus if we restore the parentheses to the abbreviated expression
$$(A \to B \to C) \to (A \to B) \to A \to C$$
we obtain the expression
$$((A \to (B \to C)) \to ((A \to B) \to (A \to C))).$$
This rule of association to the right is applied to abbreviated expressions of the form $A_1 \vee \ldots \vee A_n$, $A_1 \wedge \ldots \wedge A_n$, and $A_1 \to \ldots \to A_n$, where the A_i are any quasi-formulae. (If $n = 1$, these three expressions all stand for A_1.)

Although it is not necessary to enclose quasi-formulae of the form $s = t$ by parentheses, we often do so for sake of readability. Thus, for example, $\forall x(x = x)$ and $\neg(x = y)$ stand for the well-formed expressions $\forall xx = x$ and $\neg x = y$.

2.1   Terms and formulae

An occurrence of a variable x in any well-formed expression A is said to be a *bound occurrence* if it is an occurrence in a well-formed part of A of the form $\exists xB$, $\forall xB$, or εxB; otherwise it is a *free occurrence*. The *free variables* of A are those variables which have free occurrences in A. We denote the set of free variables in A by $V(A)$. If A is any well-formed expression and x any variable, we say x *occurs free in A within the scope of an ε-symbol* if x has a free occurrence in A within a well-formed part of A of the form εyB.

For any language $\mathscr{L}$, a quasi-formula of $\mathscr{L}$ in which no variable occurs free is called a *formula* of $\mathscr{L}$, and a quasi-term of $\mathscr{L}$ in which no variable occurs free is called a *term* of $\mathscr{L}$. Thus the sets $F_{\mathscr{L}}$ of all formulae of $\mathscr{L}$ and $T_{\mathscr{L}}$ of all terms of $\mathscr{L}$ are defined by
$$F_{\mathscr{L}} = \{A : A \text{ is a quasi-formula of } \mathscr{L} \text{ and } V(A) = \emptyset\},$$
$$T_{\mathscr{L}} = \{t : t \text{ is a quasi-term of } \mathscr{L} \text{ and } V(t) = \emptyset\}.$$
In particular, a quasi ε-term in which no variable occurs free is called an ε-*term*.

We say a quasi-term t is *free for x* in a well-formed expression A if no free occurrence of x in A is an occurrence in a well-formed part of A of the form εyB, $\exists yB$, or $\forall yB$, where y is free in t. It follows from this definition that if t is free for x in A, then on replacing all free occurrences of x in A by t, no free occurrence of a variable in t becomes bound. If t is free for x in A, we shall use the notation
$$[A]_t^x$$

to denote that (well-formed) expression which is obtained from A by replacing all free occurrences of x in A by t. We shall adopt the convention that this notation is used only when t is free for x in A. Thus when we write the formula

(1) $$\forall z([A]_z^x \leftrightarrow [B]_z^y) \to (\varepsilon x A = \varepsilon y B),$$

it is understood that z is free for x in A and for y in B. (As will be seen later, if all formulae of this form are taken as logical axioms and the usual axioms for identity are available, it is unnecessary to adopt a rule for the relabelling of bound variables.)

Since a *term* contains no free variables, then any term is free for x in A, for any x and any A. In practice, the notation $[A]_t^x$ is seldom used when t is not a term—formula (1) will be one of the few cases in which it is so used. Consequently, our convention that t is free for x in A is for the most part superfluous.

When no confusion can arise, the notation $[A]_t^x$ will often be simplified to A_t^x or just $A(t)$. However, at times the full notation is necessary. For example, $\varepsilon y[A]_t^x$ and $[\varepsilon y A]_t^x$ do not necessarily denote the same expression, and therefore the notation $\varepsilon y A_t^x$ would be ambiguous. In view of our abbreviated notation, the expressions

$$\forall x A \to A(t) \quad \text{and} \quad \neg \exists x A \to \neg A(t)$$

are to be understood as

$$\forall x A \to [A]_t^x \quad \text{and} \quad \neg \exists x A \to \neg [A]_t^x.$$

The following properties are easy to verify.

THEOREM I.1.
(i) *If s and t are terms and x and y are distinct variables, then $[[A]_s^x]_t^y$ and $[[A]_t^y]_s^x$ denote the same expression;*
(ii) *If x and y are distinct variables, then $[\exists y A]_t^x$ and $\exists y[A]_t^x$ denote the same expression, $[\forall y A]_t^x$ and $\forall y[A]_t^x$ denote the same expression, and $[\varepsilon y A]_t^x$ and $\varepsilon y[A]_t^x$ denote the same expression.*

In the case of multiple replacements of free variables, in order to economize in the use of brackets, we shall write

(1) $$[A]_{t_1 \dots t_n}^{x_1 \dots x_n}$$

instead of

(2) $$[\dots[[A]_{t_1}^{x_1}]\dots]_{t_n}^{x_n}.$$

Thus (1) denotes the expression which is obtained from A by first replacing each free occurrence of x_1 by t_1, then each free occurrence of x_2 by t_2, etc.

By Theorem I.1(i) it follows that if each t_i is a term and if the variables $x_1, \ldots, x_n$ are all distinct, then the order in which these replacements are performed is immaterial.

A *replacement operator* for a language $\mathscr{L}$ is any function, usually denoted by Σ, from a finite set of variables, called the *domain* of Σ, into $T_{\mathscr{L}}$. We shall denote the domain of Σ by *dom* Σ. For any well-formed expression A of $\mathscr{L}$ we define $[A]\Sigma$ as

$$[A]^{x_1}_{\Sigma(x_1)} \cdots {}^{x_n}_{\Sigma(x_n)}.$$

where the x_i are the distinct members of *dom* Σ. Since each $\Sigma(x_i)$ is a term, this definition is independent of the order in which the replacements are performed. If *dom* $\Sigma = \emptyset$, then $[A]\Sigma$ is A. When there is no possibility of ambiguity, we write $A\Sigma$ instead of $[A]\Sigma$.

If Σ is any replacement operator, x any variable, and t any term, then Σ^x_t is defined to be that function whose domain is *dom* $\Sigma \cup \{x\}$, such that for all $y \in dom\ \Sigma \cup \{x\}$

$$\Sigma^x_t(y) = \begin{cases} \Sigma(y) & \text{if } y \text{ is distinct from } x, \\ t & \text{if } y \text{ is } x. \end{cases}$$

For any Σ and any variable x, the *x-suppression* of Σ is that function which is obtained from Σ by restricting its domain to *dom* $\Sigma \setminus \{x\}$. Thus if Σ' is the x-suppression of Σ, then for all $y \in dom\ \Sigma \setminus \{x\}$,

$$\Sigma'(y) = \Sigma(y).$$

The following properties of replacement operators follow immediately from the definitions and Theorem I.1.

THEOREM I.2. *For any replacement operator* Σ:
  (i)   *If* A *is any quasi-formula and* $V(A) \subseteq dom\ \Sigma$, *then* $A\Sigma$ *is a formula. Similarly, for any quasi-term* t.
 (ii)   *For any well-formed expression* A, *and any variable* x, *if* $x \notin dom\ \Sigma$, *then* $[A]\Sigma^x_t$ *and* $[[A]\Sigma]^x_t$ *denote the same expression.*
(iii)   *If* Σ' *is the x-suppression of* Σ, *then* $[\exists x A]\Sigma$ *and* $\exists x[A]\Sigma'$ *denote the same expression,* $[\forall x A]\Sigma$ *and* $\forall x[A]\Sigma'$ *denote the same expression, and* $[\varepsilon x A]\Sigma$ *and* $\varepsilon x[A]\Sigma'$ *denote the same expression.*

2.2   A unifying classification of formulae

The formulae of any language $\mathscr{L}$ can be classified according to their syntactic structure as follows:

1. the atoms and the negations of atoms;
2. the formulae of the form $\neg\neg A$;
3. the formulae of the form $A \wedge B$, $\neg(A \vee B)$, and $\neg(A \to B)$;

4. the formulae of the form $A \lor B$, $\neg(A \land B)$, and $A \to B$;
5. the formulae of the form $\forall x A$ and $\neg \exists x A$;
6. the formulae of the form $\exists x A$ and $\neg \forall x A$.

Clearly, any formula of $\mathscr{L}$ is included in one and only one of the above six categories. We shall use the metalinguistic variables α, β, γ, and δ to denote arbitrary formulae of types 3, 4, 5, and 6, respectively.

In particular, a formula of type 3 is called a *conjunctive* formula. The *conjunctive components*, α_1 and α_2, of a conjunctive formula α are defined as follows: if α has the form $A \land B$, $\neg(A \lor B)$, or $\neg(A \to B)$, then α_1 and α_2 are A and B, $\neg A$ and $\neg B$, or A and $\neg B$, respectively.

A formula of type 4 is called a *disjunctive formula*. The *disjunctive components*, β_1 and β_2, of a disjunctive formula β are defined as follows: if β has the form $A \lor B$, $\neg(A \land B)$, or $A \to B$, then β_1 and β_2 are A and B, $\neg A$ and $\neg B$, or $\neg A$ and B, respectively.

A formula of type 5 is called a *universal* formula. If γ is a universal formula of the form $\forall x A$ or $\neg \exists x A$, then for any term t we use the notation $\gamma(t)$ to denote the formula $A(t)$ or $\neg A(t)$, respectively.

A formula of type 6 is called an *existential* formula. If δ is an existential formula of the form $\exists x A$ or $\neg \forall x A$, then for any term t, we use the notation $\delta(t)$ to denote the formula $A(t)$ or $\neg A(t)$, respectively. Furthermore, we use the notation $\varepsilon \delta$ to denote the term $\varepsilon x A$, if δ is $\exists x A$, and the term $\varepsilon x \neg A$, if δ is $\neg \forall x A$. Thus $\delta(\varepsilon \delta)$ denotes $A(\varepsilon x A)$ or $\neg A(\varepsilon x \neg A)$, respectively.

This system of classifying formulae into types 3, 4, 5, and 6, which is due to Smullyan [1965], greatly simplifies the metatheory since in many proofs and definitions we can avoid tiresome considerations of cases. Furthermore, since in the proofs of all our main theorems we shall not be concerned with the particular syntactic structure of a given formula, but only its more general structure as defined in the above six categories, then it is easy to see that our results hold not only for languages as defined on pages 10–12, but also for languages whose logical symbols include only one quantifier, and only one or two of the three binary connectives. In this way the above classifications and unifying notation give our metamathematical investigations an added degree of generality.

For any formula A, we define the *contrary* of A as follows: If A is the negation of some formula B, then B is the contrary of A, and if A is not a negation, then $\neg A$ is the contrary of A. Consequently, the contrary of a disjunctive formula is a conjunctive formula, and vice versa. Similarly, the contrary of an existential formula is a universal formula, and vice versa. Two formulae are said to be *contradictory* if one is the negation of the other. The following very useful properties of contradictoriness will be referred to as the *duality principle*:

1. If α and β are contradictory, then α_1 and β_1 are contradictory, and α_2 and β_2 are contradictory.
2. If γ and δ are contradictory, then for any term t, $\gamma(t)$ and $\delta(t)$ are contradictory.

2.3   The cardinality of a language

The *cardinality* of a language $\mathscr{L}$ is defined as the cardinality of its set of symbols, and by an abuse of notation will be denoted by $\overline{\overline{\mathscr{L}}}$. Since every language has enumerably many variables and individual symbols, the cardinality of a language is at least $\aleph_0$. Because we allow the vocabulary of a language to be of arbitrary cardinality, it is possible for languages to be non-enumerable. In fact, all mathematical theories can be formulated within an enumerable language; however, in recent years it has been found that non-enumerable languages play an important role in various metamathematical investigations, particularly in the realm of algebra. One metamathematical result involving non-enumerable languages which has fruitful applications is the Compactness Theorem.

In connection with our proof of the Compactness Theorem we shall make use of the following fact. If $\mathscr{L}$ is any language, and $T_{\mathscr{L}}$ is the set of its terms, then $\overline{\overline{T_{\mathscr{L}}}} \leqslant \overline{\overline{\mathscr{L}}}$. This can be proved by letting $E_{\mathscr{L}}$ be the set of expressions of $\mathscr{L}$. Since the set of symbols of $\mathscr{L}$ is infinite, and since $E_{\mathscr{L}}$ consists of all finite sequences of these symbols, then by a familiar set-theoretical argument, $\overline{\overline{E_{\mathscr{L}}}} = \overline{\overline{\mathscr{L}}}$. The result then follows from the fact that $T_{\mathscr{L}}$ is a subset of $E_{\mathscr{L}}$.

3.1   Truth functions

We begin our study of semantics by formalizing our intuitive interpretations of the symbols $\neg$ ('not'), $\vee$ ('or'), $\wedge$ ('and'), and $\rightarrow$ ('implies').

Regardless of what is meant by the words 'true' and 'false' we would like to arrange matters in such a way that the formula $\neg A$ is 'true' iff[1] A is 'false', $A \wedge B$ is 'true' iff both A and B are 'true', $A \vee B$ is 'true' iff at least one of A and B is 'true', and $A \rightarrow B$ is 'false' iff A is 'true' and B is 'false'.

We proceed as follows. Let $\mathscr{B}$ denote the set $\{0,1\}$, where the numbers 0 and 1 are called *truth values*. (Intuitively, 0 and 1 denote falsehood and truth, respectively.) An n-ary *truth function* is a function from $\mathscr{B}^n$ to $\mathscr{B}$. To the logical symbols $\neg$, $\wedge$, $\vee$, and $\rightarrow$ we assign the truth functions $H_\neg$, $H_\wedge$, $H_\vee$, and $H_\rightarrow$, respectively, where $H_\neg$ is unary and the others are binary. These functions are defined as follows. For any $m, n \in \mathscr{B}$:

$$
\begin{aligned}
H_\neg(m) &= 1 \quad \text{iff} \quad m = 0, \\
H_\wedge(m,n) &= 1 \quad \text{iff} \quad m = 1 \text{ and } n = 1, \\
H_\vee(m,n) &= 1 \quad \text{iff} \quad m = 1 \text{ or } n = 1, \\
H_\rightarrow(m,n) &= 0 \quad \text{iff} \quad m = 1 \text{ and } n = 0.
\end{aligned}
$$

[1] We use *iff* as an abbreviation for *if and only if*.

Consequently, if by some means we have assigned the truth values m and n to A and B respectively, then $H_\neg(m)$ is the correct truth value of $\neg A$, $H_\wedge(m, n)$ the correct truth value of $A \wedge B$, etc. We shall now consider one possible way in which truth values can be assigned to formulae.

3.2 Tautologies

A formula is a *molecule* if it is either an atom or a formula of the form $\forall xB$ or $\exists xB$. Thus every formula A either is a molecule or is 'built up' from molecules by means of the symbols $\wedge$, $\vee$, $\neg$, and $\rightarrow$. These molecules in A are called the *molecular constituents* of A.

A *truth assignment* is any function ψ from a finite set X of molecules into the set $\mathscr{B}$ which is such that $\psi(f) = 0$ if f is a member of X. For any formula A, if ψ is a truth assignment whose domain contains all the molecular constituents of A, then the truth value $\bar\psi(A)$ which ψ assigns to A is defined as follows by induction on the length of A:

(i) if A is a molecule, $\bar\psi(A) = \psi(A)$;
(ii) if A is of the form $\neg B$, then $\bar\psi(A) = H_\neg(\bar\psi(B))$;
(iii) if A is of the form $B * C$, where $*$ may be either $\wedge$, $\vee$, or $\rightarrow$, then $\bar\psi(A) = H_*(\bar\psi(B), \bar\psi(C))$.

A truth assignment ψ is said to be a *truth assignment for A* if the domain of ψ is the set of molecular constituents of A. Thus if A has n molecular constituents, then there are 2^n truth assignments for A. A formula A is a *tautology* if $\bar\psi(A) = 1$ for every truth assignment ψ for A, and A is a *tautological consequence* of formulae $B_1, \ldots, B_n$ if the formula $B_1 \rightarrow \ldots \rightarrow B_n \rightarrow A$ is a tautology. Thus a tautology is a formula which is 'true' no matter what truth values are assigned to its molecular constituents, and A is a tautological consequence of $B_1, \ldots, B_n$ if A is 'true' for every truth assignment which gives each of the B_i the value 'true'.

Although a tautology must certainly be regarded as a 'logically true' formula, the notions of tautology and tautological consequence do not give us the complete picture. For, in computing $\bar\psi(A)$ one looks at only the 'molecular structure' of A and disregards any occurrences of the quantifiers, ε-symbol, or identity symbol in A. For example, according to our intuitive understanding of the symbols $\exists$, $\forall$, and ε, the formula $\exists xPx$ should be a 'logical consequence' of $\forall xPx$, and the formula $P\varepsilon xPx$ should be a 'logical consequence' of $\exists xPx$. However, these are not tautological consequences since $\forall xPx \rightarrow \exists xPx$ and $\exists xPx \rightarrow P\varepsilon xPx$ are not tautologies. Thus for the type of languages which we are considering we need a more inclusive definition of 'logical consequence', i.e., one which analyzes the 'sub-atomic structure' of formulae and reflects our intuitive understanding of the symbols $\forall$, $\exists$, ε, and $=$, as well as the symbols $\neg$, $\wedge$, $\vee$, and $\rightarrow$.

Despite these inadequacies, the notion of a tautology is still a very useful one, mainly because of its decidability. For, one can effectively determine whether or not a given formula A is a tautology by computing $\bar{\psi}(A)$ for each of the 2^n truth assignments for A, where n is the number of molecular constituents of A. For this reason tautologies play an important role in the study of formal systems, as we shall see in Chapters II and III.

EXERCISES

1. Prove that any formula of the form $(A \to f) \to \neg A$ is a tautology.
2. Prove that if both A and $A \to B$ are tautologies, then B is a tautology.
3. Let α be any conjunctive formula and ψ any truth assignment for α. Prove that $\bar{\psi}(\alpha) = 1$ iff $\bar{\psi}(\alpha_1) = 1$ and $\bar{\psi}(\alpha_2) = 1$.
4. Let β be any disjunctive formula and ψ any truth assignment for β. Prove that $\bar{\psi}(\beta) = 1$ iff $\bar{\psi}(\beta_1) = 1$ or $\bar{\psi}(\beta_2) = 1$.

3.3 Models

In order to give a precise description of the semantics of our formal languages we define the notion of a model. Because of the ε-symbol, it is necessary to modify the conventional definition by equipping each model $\mathfrak{M}$ with a choice function Φ. This choice function provides a semantic interpretation of the ε-symbol.

The following familiar set-theoretic notation will be used in our definition. If M is any set and n is a positive integer, then M^n denotes the set of ordered n-tuples of M, M^{M^n} denotes the set of functions from M^n into M, and $\{0,1\}^{M^n}$ denotes the set of functions defined on M^n and taking as values 0 or 1. The set $\{0,1\}^{M^n}$ can be identified with the set of n-ary relations on M. If $n = 0$, then M^{M^n} is M, and $\{0,1\}^{M^n}$ is $\{0,1\}$.

Let $\mathscr{V}$ be any vocabulary $\langle Fn, Pr, \rho \rangle$. A *model*, $\mathfrak{M}$, for $\mathscr{V}$ is an ordered triple $\langle M, \Pi, \Phi \rangle$ which satisfies the following conditions:

1. M is a non-empty set, called the *universe* of $\mathfrak{M}$.
2. Π is a function defined on $Fn \cup Pr \cup Ind$ which assigns values in the following way:
   (i) for any individual symbol a in Ind, $\Pi(a) \in M$;
   (ii) for any n-place function symbol g in Fn, $\Pi(g) \in M^{M^n}$;
   (iii) for any n-place predicate symbol P in Pr, $\Pi(P) \in \{0,1\}^{M^n}$.
3. Φ is a choice function on M, i.e. $\Phi(N) \in N$ for any non-empty subset N of M, and $\Phi(\emptyset)$ is an arbitrary, but fixed, member of M.

The *cardinality* of a model is defined as the cardinality of its universe. Members of the universe will be denoted by the Greek letters μ and ν.

For any vocabulary $\mathscr{V}$, any model $\mathfrak{M}$ for $\mathscr{V}$, and any well-formed expression A of some language $\mathscr{L}$, we say that $\mathfrak{M}$ is *adequate* for A if A is a well-formed expression of $\mathscr{L}(\mathscr{V})$. In other words, $\mathfrak{M}$ is adequate for A if the function Π in $\mathfrak{M}$ assigns values (in the proper way) to each function and predicate symbol occurring in A.

For any model $\mathfrak{M}$, an $\mathfrak{M}$-*assignment* is any function from the set of variables Vr into the universe of $\mathfrak{M}$. $\mathfrak{M}$-assignments will be denoted by the Greek letter θ, with subscripts if necessary. If θ is an $\mathfrak{M}$-assignment, then θ_μ^x is that $\mathfrak{M}$-assignment whose value for x is μ and which otherwise coincides with θ.

If A is any well-formed expression, $\mathfrak{M}$ is any model which is adequate for A, and θ is an $\mathfrak{M}$-assignment, then the *interpretation of A with respect to $\mathfrak{M}$ and θ*, which is denoted by $\mathfrak{M}\theta(A)$, is defined as follows by induction on the length of A:

G1. *A is a variable x*: $\mathfrak{M}\theta(x) = \theta(x)$;
   A is an individual symbol a: $\mathfrak{M}\theta(a) = \Pi(a)$.
G2. *A is of the form $g^n t_1 \ldots t_n$*:
   $\mathfrak{M}\theta(g^n t_1 \ldots t_n) = \Pi(g^n)(\mathfrak{M}\theta(t_1), \ldots, \mathfrak{M}\theta(t_n))$.
G3. *A is of the form $P^n t_1 \ldots t_n$*:
   $\mathfrak{M}\theta(P^n t_1 \ldots t_n) = \Pi(P^n)(\mathfrak{M}\theta(t_1), \ldots, \mathfrak{M}\theta(t_n))$.
G4. *A is of the form $(s = t)$*:
   $\mathfrak{M}\theta(s = t) = 1$ if $\mathfrak{M}\theta(s) = \mathfrak{M}\theta(t)$, and 0 otherwise.
G5. *A is the formula f*:
   $\mathfrak{M}\theta(f) = 0$.
G6. *A is of the form $\neg B$, $(B \wedge C)$, $(B \vee C)$, or $(B \to C)$*:
   $\mathfrak{M}\theta(\neg B) \quad = H_\neg(\mathfrak{M}\theta(B))$,
   $\mathfrak{M}\theta(B \wedge C) \quad = H_\wedge(\mathfrak{M}\theta(B), \mathfrak{M}\theta(C))$,
   $\mathfrak{M}\theta(B \vee C) \quad = H_\vee(\mathfrak{M}\theta(B), \mathfrak{M}\theta(C))$,
   $\mathfrak{M}\theta(B \to C) = H_\to(\mathfrak{M}\theta(B), \mathfrak{M}\theta(C))$.
G7. *A is of the form $\exists y B$*:
   $\mathfrak{M}\theta(\exists y B) = 1$ if there exists a $\mu \in M$ such that $\mathfrak{M}\theta_\mu^y(B) = 1$; otherwise
   $\mathfrak{M}\theta(\exists y B) = 0$;
   A is of the form $\forall y B$:
   $\mathfrak{M}\theta(\forall y B) = 1$ if for all $\mu \in M$, $\mathfrak{M}\theta_\mu^y(B) = 1$; otherwise $\mathfrak{M}\theta(\forall y B) = 0$.
G8. *A is of the form $\varepsilon y B$*:
   $\mathfrak{M}\theta(\varepsilon y B) = \Phi\{\mu : \mathfrak{M}\theta_\mu^y(B) = 1\}$.

Obviously, if A is a quasi-formula, $\mathfrak{M}\theta(A) \in \{0,1\}$, and if A is a quasi-term, $\mathfrak{M}\theta(A) \in M$.

For any subset V of Vr, and any $\mathfrak{M}$-assignments θ_1 and θ_2, we write $\theta_1 \underset{V}{\sim} \theta_2$ to denote that $\theta_1(x) = \theta_2(x)$, for all $x \in V$.

THEOREM I.3. *For any well-formed expression A, any model $\mathfrak{M}$ which is adequate for A, and any $\mathfrak{M}$-assignments θ_1 and θ_2, if $\theta_1 \underset{V(A)}{\sim} \theta_2$, then $\mathfrak{M}\theta_1(A) = \mathfrak{M}\theta_2(A)$.*

Proof. The proof is by induction on the length of A. For cases G1–G6, the proof is trivial.

G7: Suppose A is of the form $\exists yB$. Since $\theta_1 \underset{V(A)}{\sim} \theta_2$, then for any $\mu \in M$, $\theta_1 {}_\mu^y \underset{V(B)}{\sim} \theta_2 {}_\mu^y$. Consequently, by the induction hypothesis, $\mathfrak{M}\theta_1 {}_\mu^y(B) = \mathfrak{M}\theta_2 {}_\mu^y(B)$, for any $\mu \in M$. It follows that there exists a $\mu \in M$ such that $M\theta_1 {}_\mu^y(B) = 1$ if and only if there exists a $\mu \in M$ such that $\mathfrak{M}\theta_2 {}_\mu^y(B) = 1$. Hence, $\mathfrak{M}\theta_1(\exists yB) = \mathfrak{M}\theta_2(\exists yB)$. Similarly, it follows that $\mathfrak{M}\theta_1(\forall yB) = \mathfrak{M}\theta_2(\forall yB)$.

G8: Suppose A is of the form εyB. As in G7, for all $\mu \in M$, $\mathfrak{M}\theta_1 {}_\mu^y(B) = \mathfrak{M}\theta_2 {}_\mu^y(B)$. Consequently, $\{\mu : \mathfrak{M}\theta_1 {}_\mu^y(B) = 1\} = \{\mu : \mathfrak{M}\theta_2 {}_\mu^y(B) = 1\}$. Hence, $\mathfrak{M}\theta_1(\varepsilon yB) = \mathfrak{M}\theta_2(\varepsilon yB)$.

COROLLARY. *If $x \notin V(A)$, then for any $\mu \in M$, $\mathfrak{M}\theta(A) = \mathfrak{M}\theta_\mu^x(A)$.*

THEOREM I.4. *For any well-formed expression A, any variable x, any quasi-term t which is free for x in A, and any model $\mathfrak{M}$ which is adequate for A and t, $\mathfrak{M}\theta([A]_t^x) = \mathfrak{M}\theta_{\mathfrak{M}\theta(t)}^x(A)$.*

Proof. If $x \notin V(A)$, then $[A]_t^x$ is A, and by the above corollary, $\mathfrak{M}\theta_{\mathfrak{M}\theta(t)}^x(A) = \mathfrak{M}\theta(A)$. Hence, in this case the theorem obviously holds. We now consider the case where $x \in V(A)$. The proof is by induction on the length of A. For cases G1–G6, the proof is trivial.

G7 and G8: Suppose A is of the form $\exists yB$, $\forall yB$, or εyB. Since $x \in V(A)$, then x and y are distinct variables. Furthermore, since t is free for x in A, then $y \notin V(t)$. Take any $\mu \in M$. By the above corollary, $\mathfrak{M}\theta_\mu^y(t) = \mathfrak{M}\theta(t)$. Hence, by the induction hypothesis

$$\mathfrak{M}\theta_\mu^y(B_t^x) = \mathfrak{M}\theta_\mu^y {}_{\mathfrak{M}\theta(t)}^x(B),$$

and therefore since x and y are distinct variables,

$$\mathfrak{M}\theta_\mu^y(B_t^x) = \mathfrak{M}\theta_{\mathfrak{M}\theta(t)}^x {}_\mu^y(B).$$

Consequently,

$$\{\mu : \mathfrak{M}\theta_\mu^y(B_t^x) = 1\} = \{\mu : \mathfrak{M}\theta_{\mathfrak{M}\theta(t)}^x {}_\mu^y(B) = 1\}.$$

This implies:

$$\mathfrak{M}\theta(\exists y[B]_t^x) = \mathfrak{M}\theta_{\mathfrak{M}\theta(t)}^x(\exists yB),$$
$$\mathfrak{M}\theta(\forall y[B]_t^x) = \mathfrak{M}\theta_{\mathfrak{M}\theta(t)}^x(\forall yB),$$
and
$$\mathfrak{M}\theta(\varepsilon y[B]_t^x) = \mathfrak{M}\theta_{\mathfrak{M}\theta(t)}^x(\varepsilon yB).$$

The theorem now follows by Theorem I.1.

If A is a formula or term, then by Theorem I.3 the interpretation of A with respect to $\mathfrak{M}$ and θ is independent of θ, since $V(A) = \emptyset$. In this case we may refer to the interpretation of A with respect to $\mathfrak{M}$ and write $\mathfrak{M}(A)$.

THEOREM I.5. *For any adequate model $\mathfrak{M}$:*

(i)   *If α is a conjunctive formula with components α_1 and α_2, then $\mathfrak{M}(\alpha) = 1$ iff $\mathfrak{M}(\alpha_1) = 1$ and $\mathfrak{M}(\alpha_2) = 1$.*

(ii)   *For any formulae A and B, $\mathfrak{M}(A \leftrightarrow B) = 1$ iff $\mathfrak{M}(A) = \mathfrak{M}(B)$.*

(iii)   *For any universal formula γ and any term t, if $\mathfrak{M}(\gamma) = 1$, then $\mathfrak{M}(\gamma(t)) = 1$.*

(iv)   *For any existential formula δ, if $\mathfrak{M}(\delta) = 1$, then $\mathfrak{M}(\delta(\varepsilon\delta)) = 1$.*

(v)   *If $\mathfrak{M}(\forall x(A \leftrightarrow B)) = 1$, then $\mathfrak{M}(\varepsilon x A) = \mathfrak{M}(\varepsilon x B)$.*

(vi)   *For any ε-term $\varepsilon x A$, $\mathfrak{M}(\varepsilon x A) = \mathfrak{M}(\varepsilon y[A]_y^x)$.*

(vii)   *For any terms s and t, if $\mathfrak{M}(s) = \mathfrak{M}(t)$ and $\mathfrak{M}(A_s^x) = 1$, then $\mathfrak{M}(A_t^x) = 1$.*

Proof. The proofs of (i) and (ii) follow trivially from our truth functional interpretations of $\neg$, $\wedge$, $\vee$, and $\rightarrow$, and from the definition of $\leftrightarrow$.

(iii): Suppose γ is of the form $\forall x A$. Let θ be any $\mathfrak{M}$-assignment. Since $\mathfrak{M}(\forall x A) = 1$, then $\mathfrak{M}\theta(\forall x A) = 1$, and for all $\mu \in M$, $\mathfrak{M}\theta_\mu^x(A) = 1$. In particular, $\mathfrak{M}\theta_{\mathfrak{M}\theta(t)}^x(A) = 1$, and by Theorem I.4 $\mathfrak{M}\theta(A(t)) = 1$. Hence, $\mathfrak{M}(\gamma(t)) = 1$. The proof is similar for the case where γ is of the form $\neg\exists x A$.

(iv): Suppose δ is of the form $\neg\forall x A$. Let θ be any $\mathfrak{M}$-assignment. Since $\mathfrak{M}(\delta) = 1$, then $\mathfrak{M}\theta(\neg\forall x A) = 1$, and therefore $\mathfrak{M}\theta(\forall x A) = 0$. Hence there exists a $\mu \in M$ such that $\mathfrak{M}\theta_\mu^x(A) = 0$. Consequently, $\mathfrak{M}\theta_\mu^x(\neg A) = 1$. Let $N = \{\mu : \mathfrak{M}\theta_\mu^x(\neg A) = 1\}$. Since $N \neq \emptyset$, $\Phi(N) \in N$. But $\mathfrak{M}\theta(\varepsilon x \neg A) = \Phi(N)$. Therefore, $\mathfrak{M}\theta_{\mathfrak{M}\theta(\varepsilon x \neg A)}^x(\neg A) = 1$, and by Theorem I.4 $\mathfrak{M}\theta(\neg A(\varepsilon x \neg A)) = 1$. Consequently, $\mathfrak{M}(\delta(\varepsilon\delta)) = 1$. The proof is similar for the case where δ is of the form $\exists x A$.

(v): Let θ be any $\mathfrak{M}$-assignment. Since $\mathfrak{M}(\forall x(A \leftrightarrow B)) = 1$, then for all $\mu \in M$, $\mathfrak{M}\theta_\mu^x(A \leftrightarrow B) = 1$. Hence by (ii), for all $\mu \in M$, $\mathfrak{M}\theta_\mu^x(A) = \mathfrak{M}\theta_\mu^x(B)$. Thus $\{\mu : \mathfrak{M}\theta_\mu^x(A) = 1\} = \{\mu : \mathfrak{M}\theta_\mu^x(B) = 1\}$, and it follows that $\mathfrak{M}(\varepsilon x A) = \mathfrak{M}(\varepsilon x B)$.

(vi): If x and y are the same variable, the proof is trivial. Suppose x and y are distinct variables. Since $\varepsilon x A$ is a term, then $y \notin V(A)$. Furthermore, by our convention (p. 13), y is free for x in A. Let $N_1 = \{\mu : \mathfrak{M}\theta_\mu^x(A) = 1\}$ and let $N_2 = \{\mu : \mathfrak{M}\theta_\mu^y(A_y^x) = 1\}$. It will be sufficient to prove $N_1 = N_2$. Since $\theta_\mu^y(y) = \mu$, then by Theorem I.4

$$\begin{aligned}
\mathfrak{M}\theta_\mu^y(A_y^x) &= \mathfrak{M}\theta_{\mu}^{y}{}_{\mu}^{x}(A), \\
&= \mathfrak{M}\theta_{\mu}^{x}{}_{\mu}^{y}(A), \\
&= \mathfrak{M}\theta_\mu^x(A) \quad \text{since} \quad y \notin V(A).
\end{aligned}$$

Consequently $N_1 = N_2$.

(vii): The proof follows immediately from Theorem I.4.

EXERCISES

1. Prove that for any formula A, if $\mathfrak{M}$ is adequate for A and A is a tautology, then $\mathfrak{M}(A) = 1$.

2. Prove that, if $\mathfrak{M}(A) = 1$ and $\mathfrak{M}(A \rightarrow B) = 1$, then $\mathfrak{M}(B) = 1$.

3.4   Satisfiability and the semantic consequence relation

Let A be a formula of some language $\mathscr{L}$ and $\mathfrak{M}$ any model. We say $\mathfrak{M}$ *satisfies* A, or A is *true* in $\mathfrak{M}$, if $\mathfrak{M}$ is adequate for A and $\mathfrak{M}(A) = 1$. We write $\mathfrak{M}$ *Sat* A to denote that $\mathfrak{M}$ satisfies A. If X is any set of formulae, we say $\mathfrak{M}$ *satisfies* X, denoted by $\mathfrak{M}$ *Sat* X, if $\mathfrak{M}$ satisfies all the members of X. We say X is *satisfiable* or *has a model*, denoted by *Sat* X, if there exists a model which satisfies X, and we say X is $\mathfrak{m}$-*satisfiable*, denoted by $\mathfrak{m}$-*Sat* X, if there exists a model of cardinality $\mathfrak{m}$ which satisfies X. For any set of formulae X and any formula A, we say A is a *semantic consequence* of X, denoted by $X \vDash A$, if every model which is adequate for A and which satisfies X also satisfies A. Finally A is *valid*, denoted by $\vDash A$, if A is a semantic consequence of the null set. It follows that A is valid if and only if every model which is adequate for A satisfies A.

For any language $\mathscr{L}$, we define the *finitary semantic closure for* $\mathscr{L}$ as the function, denoted by C_s, such that for every $X \subseteq F_{\mathscr{L}}$:

$$C_s(X) = \{A : A \in F_{\mathscr{L}} \text{ and there exists a finite subset } Y \text{ of } X \text{ such that } Y \vDash A\}.$$

Clearly, if $A \in C_s(X)$, then $X \vDash A$. For finite X, the converse is also true. The fact that the converse is true for infinite X will follow from the Compactness Theorem (p. 29).

For convenience of notation we shall write $C_s(X, A)$ instead of $C_s(X \cup \{A\})$, where X is a set of formulae and A is a formula. The following theorem follows readily from the definition of C_s.

THEOREM I.6. *For any $X \subseteq F_{\mathscr{L}}$:*
  (i) $X \subseteq C_s(X)$;
 (ii) $C_s(C_s(X)) \subseteq C_s(X)$;
(iii)   *if $Y \subseteq X$, then $C_s(Y) \subseteq C_s(X)$;*
(iv)   *if $A \in X$, then there exists a finite $Y \subseteq X$ such that $A \in C_s(Y)$.*

The proof is left as an exercise.

THEOREM I.7. *If $f \in C_s(X)$, then there exists a finite subset Y of X which is unsatisfiable.*

Proof. Since $f \in C_s(X)$, then there exists some finite subset Y of X such that $Y \vDash f$. But for any model $\mathfrak{M}$, $\mathfrak{M}(f) = 0$. Hence there is no model which satisfies Y.

In view of Theorem I.7, to prove the Compactness Theorem it will be sufficient to prove that for any $X \subseteq F_{\mathscr{L}}$:

(1)   if $f \notin C_s(X)$, then $\mathfrak{m}$-*Sat* X, for some $\mathfrak{m} \leqslant \overline{\overline{\mathscr{L}}}$.

This assertion follows as a special case from the main theorem to be proved in §4, Theorem I.11. The following theorem will be used in connection with that result.

THEOREM I.8. *Let C_s be the finitary semantic closure for some language $\mathcal{L}$. Then for any $X \subseteq F_{\mathcal{L}}$, any $A \in F_{\mathcal{L}}$, any $\alpha, \gamma, \delta \in F_{\mathcal{L}}$, and any $t \in T_{\mathcal{L}}$:*

(i) $A \in C_s(X)$ *iff* $f \in C_s(X, \neg A)$;

(ii) $\alpha \in C_s(X)$ *iff* $\alpha_1, \alpha_2 \in C_s(X)$;

(iii)   *if* $\gamma \in X$, *then* $\gamma(t) \in C_s(X)$;

(iv)   *if* $\delta \in X$, *then* $\delta(\varepsilon\delta) \in C_s(X)$;

(v)   *if* $(s = t) \in X$ *and* $B_s^x \in X$, *then* $B_t^x \in C_s(X)$;

(vi)   *for any* ε-*terms* $\varepsilon x A$ *and* $\varepsilon y B$ *of* $\mathcal{L}$, *if* $\forall z(A_z^x \leftrightarrow B_z^y) \in X$, *then* $(\varepsilon x A = \varepsilon y B) \in C_s(X)$.

Proof. (i): Assume $A \in C_s(X)$. Then $Y \vDash A$ for some finite $Y \subseteq X$. Let $Z = Y \cup \{\neg A\}$. Clearly, Z is unsatisfiable; hence, $Z \vDash f$. Since $Z \subseteq X \cup \{\neg A\}$, then $f \in C_s(X, \neg A)$. Conversely, assume that $f \in C_s(X, \neg A)$. By Theorem I.7, there exists a finite subset Y of $X \cup \{\neg A\}$ which is unsatisfiable. Let $Z = Y \setminus \{\neg A\}$. It is enough to prove that $Z \vDash A$. Assume the contrary, i.e. assume that there exists a model $\mathfrak{M}$ such that $\mathfrak{M}$ *Sat* Z, but $\mathfrak{M}(A) = 0$. Then $\mathfrak{M}(\neg A) = 1$ and $\mathfrak{M}$ *Sat* Y, which is impossible.

The proofs of (ii)–(vi) follow by Theorem I.5.

4   The Satisfiability Theorem

In this section we prove our main result, the Satisfiability Theorem, from which the Compactness Theorem follows as a special case. This abstract result is obtained by introducing the general notion of a logical closure C and then proving that for any set X of formulae of $\mathcal{L}$:

$$\text{if } f \notin C(X), \text{ then } \mathfrak{m}\text{-}Sat\ X, \text{ for some } \mathfrak{m} \leqslant \overline{\overline{\mathcal{L}}}.$$

We first define the familiar notion of a *finitary closure operation* C on any set S as a function from the power set of S into itself such that, for any $X \subseteq S$, the following conditions hold:

C1. $X \subseteq C(X)$;

C2. $C(C(X)) \subseteq C(X)$;

C3. if $Y \subseteq X$, then $C(Y) \subseteq C(X)$;

C4. if $A \in C(X)$, then there exists a finite $Y \subseteq X$ such that $A \in C(Y)$.

The notion of a finitary closure operation, which is fundamental to both algebra and logic, was apparently first applied to logic by Tarski [1930].

If $\mathcal{L}$ is a language and C is a finitary closure operation on $F_{\mathcal{L}}$, then for any $X \subseteq F_{\mathcal{L}}$, we say X is *maximal under* C, denoted by $Max_C X$, if:

1. $f \notin C(X)$;
2. for all $Y \subseteq F_{\mathscr{L}}$, if $X \subseteq Y$ and $f \notin C(Y)$, then $X = Y$.

THEOREM I.9. *If C is a finitary closure operation on $F_{\mathscr{L}}$ and $Max_C X$, then* $C(X) = X$.

Proof. By C1, $X \subseteq C(X)$. Since $Max_C X$, then $f \notin C(X)$, and hence, by C2, $f \notin C(C(X))$. But this implies $X = C(X)$ by the maximality of X.

THEOREM I.10. *If $f \notin C(X)$, then there exists some $Y \subseteq F_{\mathscr{L}}$, such that $X \subseteq Y$ and $Max_C Y$.*

Proof. Let $\mathfrak{S} = \{X' : X \subseteq X', X' \subseteq F_{\mathscr{L}}, \text{ and } f \notin C(X')\}$. We shall use Zorn's lemma to prove that $\mathfrak{S}$ contains maximal elements. Let $\mathfrak{T}$ be any subset of $\mathfrak{S}$ which is totally ordered by inclusion, and let $Y = \bigcup_{X' \in \mathfrak{T}} X'$. In order to apply Zorn's lemma we need only show that $Y \in \mathfrak{S}$. Clearly $Y \subseteq F_{\mathscr{L}}$ and $X \subseteq Y$. To prove $f \notin C(Y)$ we assume the contrary and produce a contradiction. If $f \in C(Y)$, then by C4, $f \in C(Y_0)$ for some finite subset Y_0 of Y. Since Y_0 is finite, then $Y_0 \subseteq X'$ for some $X' \in \mathfrak{T}$, and by C3, $C(Y_0) \subseteq C(X')$. Hence $f \in C(X')$, which contradicts the definition of $\mathfrak{T}$. This completes the proof.

Note: If the vocabulary of $\mathscr{L}$ is finite or denumerable, then Zorn's lemma (and the axiom of choice) can be avoided in the usual way by enumerating the formulae in $F_{\mathscr{L}}$.

Let $\mathscr{L}$ be any formal language. A finitary closure operation C on $F_{\mathscr{L}}$ which possesses the following six properties is called a *logical closure* for $\mathscr{L}$: For any $X \subseteq F_{\mathscr{L}}$, any A, α, γ, and $\delta \in F_{\mathscr{L}}$, and any s and $t \in T_{\mathscr{L}}$:

L1. $A \in C(X)$ iff $f \in C(X, \neg A)$.
L2. $\alpha \in C(X)$ iff $\alpha_1, \alpha_2 \in C(X)$.
L3. If $\gamma \in X$, then $\gamma(t) \in C(X)$.
L4. If $\delta \in X$, then $\delta(\varepsilon\delta) \in C(X)$.
L5. For any atom B and any variable x which does not occur free in B within the scope of an ε-symbol, if $s = t \in X$ and $B_s^x \in X$, then $B_t^x \in C(X)$.
L6. For any ε-terms $\varepsilon x A$ and $\varepsilon y B$, if $\forall z(A_z^x \leftrightarrow B_z^y) \in X$, then $(\varepsilon x A = \varepsilon y B) \in C(X)$.

For example, by Theorems I.6 and I.8, the finitary semantic closure, C_s, is a logical closure.

THEOREM I.11 (The Satisfiability Theorem). *Let C be a logical closure for a language $\mathscr{L}$. For any $X \subseteq F_{\mathscr{L}}$, if $f \notin C(X)$, then $\mathfrak{m}\text{-}Sat\ X$, for some $\mathfrak{m} \leqslant \overline{\overline{\mathscr{L}}}$.*

By Theorem I.10, if $f \notin C(X)$, then X is contained in some set Y which is maximal under C. Also, if $X \subseteq Y$, then $\mathfrak{m}$-*Sat* Y implies $\mathfrak{m}$-*Sat* X. Consequently, to prove Theorem I.11, it will be sufficient to prove that:

$$Max_C X \text{ implies } \mathfrak{m}\text{-}Sat\ X, \text{ for some } \mathfrak{m} \leqslant \overline{\overline{\mathscr{L}}}.$$

Throughout the following lemmas and definitions it is assumed that X *is some fixed set which is maximal under the logical closure* C. Thus, by Theorem I.9, X is closed, i.e. $C(X) = X$. The following lemmas and definitions will enable us to construct the required model. The basic idea of this proof is in the spirit of Henkin's proof of the completeness of the predicate calculus (cf. Henkin [1949]), since the notion of a maximal set under C is a generalization of his notion of a 'maximal consistent set'. Although the presence of the ε-symbol has a complicating effect in some respects, in other respects this symbol simplifies the proof since the availability of ε-terms makes it unnecessary to adjoin new constant symbols to the language $\mathscr{L}$, as is done by Henkin.

Some of the techniques used here are adaptations of methods used by Hermes [1965] in his proof of the completeness of his *Termlogik mit Auswahloperator*. The proof given here is an improvement on Hermes' result, however, since by its abstract nature our proof does not depend on a particular set of logical axioms and rules of inference. Furthermore, Hermes' result holds only for languages with a denumerable set of symbols, whereas our result holds for languages of arbitrary infinite cardinality.

LEMMA 1.

(i) $A \in X$ *iff* $\neg A \notin X$;
(ii) $A \to B \in X$ *iff* $A \notin X$ *or* $B \in X$;
(iii) $A \vee B \in X$ *iff* $A \in X$ *or* $B \in X$;
(iv) $A \wedge B \in X$ *iff* $A \in X$ *and* $B \in X$;
(v) $A \leftrightarrow B \in X$ *iff* $(A \in X$ *iff* $B \in X)$;
(vi) $\exists x A \in X$ *iff* $A(t) \in X$, *for some* $t \in T_{\mathscr{L}}$;
(vii) $\forall x A \in X$ *iff* $A(t) \in X$, *for all* $t \in T_{\mathscr{L}}$;
(viii) *if* $A(\varepsilon x \neg A) \in X$, *then* $\forall x A \in X$;
(ix) $(t = t) \in X$, *for all* $t \in T_{\mathscr{L}}$.

Proof.
(i): $A \in X$ iff $\mathit{f} \in C(X, \neg A)$ by L1,
         iff $\neg A \notin X$ by $Max_C X$.
(ii): $A \to B \notin X$ iff $\neg(A \to B) \in X$ by (i),
         iff $A \in X$ and $\neg B \in X$ by L2,
         iff $A \in X$ and $B \notin X$ by (i).
(iii): The proof is similar to that of (ii).
(iv): The proof follows immediately by L2.

(v): The proof is immediate by the definition of $\leftrightarrow$, using (ii) and (iv).

(vi): Assume $\exists x A \in X$. By L4, $A(\varepsilon x A) \in X$. Hence there exists a $t \in T_{\mathscr{L}}$, such that $A(t) \in X$. Conversely, assume $A(t) \in X$, for some $t \in T_{\mathscr{L}}$. By (i), $\neg A(t) \notin X$. Consequently, by (the contrapositive of) L3, $\neg \exists x A \notin X$. Hence, by (i) again, $\exists x A \in X$.

(vii): The proof is similar to that of (vi).

(viii): If $A(\varepsilon x \neg A) \in X$, then by (i), $\neg A(\varepsilon x \neg A) \notin X$. Consequently, by (the contrapositive of) L4, $\neg \forall x A \notin X$. Hence by (i), $\forall x A \in X$.

(ix): Let t_0 be the term $\varepsilon x \neg (\neg (x = x) \leftrightarrow \neg (x = x))$. By (v) we have $\neg (t_0 = t_0) \leftrightarrow \neg (t_0 = t_0) \in X$. Thus, by (viii) and the definition of t_0, $\forall x (\neg (x = x) \leftrightarrow \neg (x = x)) \in X$. So, by L6, $(\varepsilon x \neg (x = x) = \varepsilon x \neg (x = x)) \in X$. Again by (viii), $\forall x (x = x) \in X$. Consequently, by (vii), for any $t \in T_{\mathscr{L}}$, $(t = t) \in X$.

We now define a binary relation $\sim$ on $T_{\mathscr{L}}$ as follows: for any s, t, $\in T_{\mathscr{L}}$, $s \sim t$ iff $(s = t) \in X$.

LEMMA 2. The relation $\sim$ is an equivalence relation on $T_{\mathscr{L}}$.

Proof. (i): That $\sim$ is reflexive follows from Lemma 1(ix).

(ii): The symmetry of $\sim$ is proved as follows. Assume $s \sim t$. Thus $(s = t) \in X$. Let B be the quasi-formula $(x = s)$. Thus $B_s^x \in X$, by Lemma 1(ix). Hence, by L5, $B_t^x \in X$, i.e. $(t = s) \in X$. Therefore, $t \sim s$.

(iii): To prove that $\sim$ is transitive, assume $r \sim s$ and $s \sim t$. Thus $(s = t) \in X$, and if B is the quasi-formula $(r = x)$, then $B_s^x \in X$. Thus, by L5, $B_t^x \in X$, i.e. $(r = t) \in X$, and so $r \sim t$.

LEMMA 3. *If g^n and P^n are n-place function and predicate symbols, respectively, of $\mathscr{L}$, and $s_i \sim t_i$ for each $i = 1, \ldots, n$, then*
(i) $g^n s_1 \ldots s_n \sim g^n t_1 \ldots t_n$,
(ii) $P^n s_1 \ldots s_n \in X$ iff $P^n t_1 \ldots t_n \in X$.

Proof. (i): For each $i = 1, \ldots, n$, let A_i be the quasi-formula

$$(g^n s_1 \ldots s_n = g^n t_1 \ldots t_{i-1} x s_{i+1} \ldots s_n).$$

Thus $A_{1 \, s_1}^{\,\,\, x}$ is the formula

$$(g^n s_1 \ldots s_n = g^n s_1 \ldots s_n).$$

Consequently, by Lemma 1(ix), $A_{1 \, s_1}^{\,\,\, x} \in X$, and by L5, $A_{1 \, t_1}^{\,\,\, x} \in X$. However, for each $i = 1, \ldots, n$, $A_{i \, t_i}^{\,\,\, x}$ and $A_{i+1 \, s_{i+1}}^{\,\,\,\,\,\,\,\, x}$ are the same formula. Therefore, by n applications of L5, we obtain $A_{n \, t_n}^{\,\,\, x} \in X$, i.e. $(g^n s_1 \ldots s_n = g^n t_1 \ldots t_n) \in X$, and so $g^n s_1 \ldots s_n \sim g^n t_1 \ldots t_n$.

(ii): The proof is similar to that of (i) except that we let A_i be the quasi-formula $P t_1 \ldots t_{i-1} x s_{i+1} \ldots s_n$.

We now let $|t|$ denote the equivalence class of any term t, and let M denote the set of equivalence classes on $T_{\mathscr{L}}$. The following definition is due to Hermes [1965].

For any subset N of M, we say N is *representable* if there exists a quasi-formula A of $\mathscr{L}$ and a variable x, such that for all $t \in T_{\mathscr{L}}$, $|t| \in N$ iff $A_t^x \in X$. In this case we say $\langle A,x \rangle$ is a *representative* of N. Clearly, if $\langle A,x \rangle$ represents N, then no variable other than x occurs free in A, and therefore $\varepsilon x A$ is a term.

LEMMA 4. *If $\langle A,x \rangle$ and $\langle B,y \rangle$ are both representatives of N, then $\varepsilon x A \sim \varepsilon y B$.*

Proof. Let z be any variable not appearing (either free or bound) in A or B. Let A' be A_z^x and B' be B_z^y. Thus for any term t, $A'{}_t^z$ is A_t^x, and $B'{}_t^z$ is B_t^y. Consequently $\langle A',z \rangle$ and $\langle B',z \rangle$ are representatives of N. Let t_0 be the term $\varepsilon z \neg (A' \leftrightarrow B')$. Since $\langle A',z \rangle$ and $\langle B',z \rangle$ represent N, then $|t_0| \in N$ iff $A'{}_{t_0}^z \in X$ and $|t_0| \in N$ iff $B'{}_{t_0}^z \in X$. Thus $A'{}_{t_0}^z \in X$ iff $B'{}_{t_0}^z \in X$, and by Lemma 1(v), $A'{}_{t_0}^z \leftrightarrow B'{}_{t_0}^z \in X$. By Lemma 1(viii) and our choice of t_0, this implies that $\forall z (A' \leftrightarrow B') \in X$. Consequently, by L6, $(\varepsilon x A = \varepsilon y B) \in X$, and $\varepsilon x A \sim \varepsilon y B$.

LEMMA 5. *If $\langle A,x \rangle$ is a representative of a non-empty set N, then $|\varepsilon x A| \in N$.*

Proof. Since N is non-empty and $\langle A,x \rangle$ represents N, then there exists a term t such that $A(t) \in X$. Thus by Lemma 1(vi), $\exists x A \in X$, and by L4, $A(\varepsilon x A) \in X$. Hence $|\varepsilon x A| \in N$.

The model $\mathfrak{M}$ for the set X that we need in order to prove Theorem I.11 is now defined in the following way.

1. The universe of $\mathfrak{M}$ is the set M above, i.e. the set of equivalence classes on $T_{\mathscr{L}}$ under the equivalence relation $\sim$.
2. The function Π is defined as follows:
   (i) For any individual symbol a, $\Pi(a) = |a|$.
   (ii) For any g^n in the vocabulary of $\mathscr{L}$ and any $|t_1|, \ldots, |t_n|$ in M, $\Pi(g^n)(|t_1|, \ldots, |t_n|) = |g^n t_1 \ldots t_n|$. By Lemma 3, this definition is independent of the choice of the t_i.
   (iii) For any P^n in the vocabulary of $\mathscr{L}$ and any $|t_1|, \ldots, |t_n|$ in M, $\Pi(P^n)(|t_1|, \ldots, |t_n|) = 1$ iff $P^n t_1 \ldots t_n \in X$. Again by Lemma 3, this definition is independent of the choice of the t_i.
3. The choice function Φ is defined as follows for any $N \subseteq M$.
   Case 1: If N is representable, $\Phi(N) = |\varepsilon x A|$, where $\langle A,x \rangle$ is one of its representatives. By Lemma 4, this definition is independent of the choice of representative.
   Case 2: If N is not representable, (and hence $N \neq \emptyset$, since the null set is represented by $\langle \neg(x = x),x \rangle$), then $\Phi(N)$ is an arbitrary member of N.

By Lemma 5, if $N \neq \emptyset$, then $\Phi(N) \in N$. Thus under this definition Φ is a choice function on M.

Under these definitions of M, Π, and Φ, the triple $\langle M, \Pi, \Phi \rangle$, which we shall denote by $\mathfrak{M}$, is a model for the vocabulary of $\mathscr{L}$. Furthermore, since its universe, M, is $T_{\mathscr{L}}/\sim$, then $\overline{\overline{M}} \leqslant \overline{\overline{T}}_{\mathscr{L}}$. Therefore the cardinality of the model $\mathfrak{M}$ is less than or equal to the cardinality of the language $\mathscr{L}$ (see page 16). All that remains to prove is that $\mathfrak{M}$ *Sat X*.

Let θ be any $\mathfrak{M}$-assignment and Σ any replacement operator for the language $\mathscr{L}$. Since for any $x \in dom\ \Sigma$, $\Sigma(x) \in T_{\mathscr{L}}$, then $|\Sigma(x)|$ is defined for $x \in dom\ \Sigma$. Let V be any subset of the set of variables, Vr. We write $\theta \underset{V}{\simeq} \Sigma$ to denote that $V \subseteq dom\ \Sigma$ and for all $x \in V$, $\theta(x) = |\Sigma(x)|$.

LEMMA 6. *Let A be any well-formed expression of $\mathscr{L}$. If θ is any $\mathfrak{M}$-assignment and Σ is any replacement operator such that $\theta \underset{V(A)}{\simeq} \Sigma$, then:*

(i) $\mathfrak{M}\theta(A) = |A\Sigma|$, *if A is a quasi-term;*
(ii) $\mathfrak{M}\theta(A) = 1$ *iff $A\Sigma \in X$, if A is a quasi-formula.*

Proof. The proof is by induction on the length of A.

G1. *A is a variable x:* $\mathfrak{M}\theta(x) = \theta(x) = |\Sigma(x)| = |x\Sigma|$;
A is an individual symbol a: $\mathfrak{M}\theta(a) = \Pi(a) = |a| = |a\Sigma|$.

G2. *A is of the form $g^n t_1 \ldots t_n$:* Since for each $i = 1, \ldots, n$, $V(t_1) \subseteq V(A)$, then $\theta \underset{V(t_i)}{\simeq} \Sigma$. Therefore, by the induction hypothesis, $\mathfrak{M}\theta(t_i) = |t_i\Sigma|$. Now,

$$\mathfrak{M}\theta(g^n t_1 \ldots t_n) = \Pi(g^n)(\mathfrak{M}\theta(t_1), \ldots, \mathfrak{M}\theta(t_n)),$$
$$= \Pi(g^n)(|t_1\Sigma|, \ldots, |t_n\Sigma|) \quad \text{by induction hypothesis,}$$
$$= |g^n[t_1]\Sigma \ldots [t_n]\Sigma| \quad \text{by the definition of } \Pi,$$
$$= |[g^n t_1 \ldots t_n]\Sigma|.$$

G3. *A is of the form $P^n t_1 \ldots t_n$:*
$$\mathfrak{M}\theta(P^n t_1 \ldots t_n) = 1 \quad \text{iff} \quad \Pi(P^n)(\mathfrak{M}\theta(t_1), \ldots, \mathfrak{M}\theta(t_n)) = 1,$$
$$\text{iff} \quad \Pi(P^n)(|t_1\Sigma|, \ldots, |t_n\Sigma|) = 1 \quad \text{by induction hypothesis,}$$
$$\text{iff} \quad P^n[t_1]\Sigma \ldots [t_n]\Sigma \in X \quad \text{by the definition of } \Pi,$$
$$\text{iff} \quad [P^n t_1 \ldots t_n]\Sigma \in X.$$

G4. *A is of the form $(s = t)$:*
$$\mathfrak{M}\theta(s = t) = 1 \quad \text{iff} \quad \mathfrak{M}\theta(s) = \mathfrak{M}\theta(t)$$
$$\text{iff} \quad |s\Sigma| = |t\Sigma| \quad \text{by induction hypothesis,}$$
$$\text{iff} \quad (s\Sigma = t\Sigma) \in X \quad \text{by the definition of } \sim,$$
$$\text{iff} \quad [(s = t)]\Sigma \in X.$$

G5. *A is the formula f:* $\mathfrak{M}\theta(f) = 0$, and by $Max_C X$, $f \notin X$. Consequently, $\mathfrak{M}\theta(f) = 1$ iff $f\Sigma \in X$, since both sides of this equivalence are false.

G6. *A is of the form* $\neg B$, $(B \to C)$, $(B \lor C)$, *or* $(B \land C)$: The proof follows immediately by Lemma 1 (i)–(iv).

G7–G8. *A is of the form* $\exists x B$, $\forall x B$, $\varepsilon x B$: Let Σ' be the x-suppression of Σ (see p. 14). Let $N = \{\mu : \mathfrak{M}\theta_\mu^x(B) = 1\}$. Since $\theta \underset{V(A)}{\simeq} \Sigma$, then for any $t \in T_{\mathscr{L}}$ $\theta_{|t|}^x \underset{V(B)}{\simeq} \Sigma' \, {}_t^x$. We first prove that (1) for all $t \in T_{\mathscr{L}}$, $|t| \in N$ iff $[B\Sigma']_t^x \in X$.

For any $t \in T_{\mathscr{L}}$, $|t| \in N$ iff $\mathfrak{M}\theta_{|t|}^x(B) = 1$ by definition of N,
iff $[B]\Sigma' \, {}_t^x \in X$ by induction hypothesis,
iff $[B\Sigma']_t^x \in X$ Theorem I.2(ii) (p. 14).

Consequently $\langle B\Sigma', x \rangle$ is a representative of N, as required.

Case 1. *A is of the form* $\exists x B$:
$\mathfrak{M}\theta(\exists x B) = 1$ iff $N \neq \emptyset$
iff there exists a $t \in T_{\mathscr{L}}$ such that $[B\Sigma']_t^x \in X$, by (1),
iff $\exists x [B]\Sigma' \in X$ by Lemma 1(vi),
iff $[\exists x B]\Sigma \in X$ by Theorem I.2(iii) (p. 14).

Case 2. *A is of the form* $\forall x B$:
$\mathfrak{M}\theta(\forall x B) = 1$ iff $N = M$,
iff for all $t \in T_{\mathscr{L}}$, $[B\Sigma']_t^x \in X$, by (1),
iff $\forall x [B]\Sigma' \in X$ by Lemma 1(vii),
iff $[\forall x B]\Sigma \in X$ by Theorem I.2(iii).

Case 3. *A is of the form* $\varepsilon x B$:
$\mathfrak{M}\theta(\varepsilon x B) = \Phi(N)$
$= \left| \varepsilon x [B]\Sigma' \right|$ by definition of Φ, since $\langle B\Sigma', x \rangle$ represents N,
$= \left\| [\varepsilon x B]\Sigma \right|$ by Theorem I.2(iii).

This completes the proof of Lemma 6.

Using Lemma 6, we now prove that for any $A \in F_{\mathscr{L}}$, $\mathfrak{M}(A) = 1$ iff $A \in X$. Take any $\mathfrak{M}$-assignment θ and any replacement operator Σ. Since A is a formula, then $V(A) = \emptyset$, and therefore $\theta \underset{V(A)}{\simeq} \Sigma$. Thus by Lemma 6, $\mathfrak{M}\theta(A) = 1$ iff $A\Sigma \in X$. But $A\Sigma$ is A itself, and $\mathfrak{M}\theta(A) = \mathfrak{M}(A)$. Therefore, $\mathfrak{M}(A) = 1$ iff $A \in X$. Consequently, $\mathfrak{M}$ *Sat* X. This completes the proof of Theorem I.11.

THEOREM I.12 (The Compactness Theorem). *If X is a set of formulae of some language $\mathscr{L}$ such that every finite subset of X is satisfiable, then X is m-satisfiable, for some* $\mathfrak{m} \leqslant \bar{\mathscr{L}}$.

Proof. By Theorems I.6 and I.8, C_s is a logical closure for $\mathscr{L}$. By Theorem I.7, if every finite subset of X is satisfiable, then $f \notin C_s(X)$. The theorem, therefore, follows from Theorem I.11.

THEOREM I.13 (Löwenheim-Skolem Theorem). *If X is any satisfiable set of formulae of some language $\mathscr{L}$, then X is $\mathfrak{m}$-satisfiable, for some $\mathfrak{m} \leqslant \overline{\overline{\mathscr{L}}}$.*

Proof. If X is satisfiable, then every finite subset of X is satisfiable. The result then follows from Theorem I.12.

We can strengthen the statements of Theorems I.12 and I.13 by observing that no condition is imposed on the language $\mathscr{L}$ except that X is a set of its formulae. Thus we may consider $\mathscr{L}$ to be the language whose vocabulary consists of just those function symbols and predicate symbols which appear in the members of X. In this case, $\overline{\overline{\mathscr{L}}} = \max\{\aleph_0, \overline{X}\}$, and we can state the conclusions of Theorems I.12 and I.13 as 'X is $\mathfrak{m}$-satisfiable, for some $\mathfrak{m} \leqslant \max\{\aleph_0, \overline{X}\}$'.

Consequently, using the abstract notion of a logical closure, we have managed to prove the Compactness Theorem for a formal language $\mathscr{L}$ without becoming involved in a particular deductive structure for that language, i.e. without choosing some particular set of logical axioms and rules of inference and then first proving the Completeness Theorem. Another advantage of this abstract approach is that when we do impose a particular deductive structure on the language $\mathscr{L}$, by specifying certain axioms and rules of inference, we can then prove the completeness of this formal system merely by showing that its deductive closure is a logical closure.

There are other ways of proving the Compactness Theorem without first proving the Completeness Theorem. One such method involves the use of ultraproducts (cf. Frayne, Morel, and Scott [1962]).

EXERCISES

1. Prove that if $X \vDash A$, then there exists a finite $Y \subseteq X$ such that $Y \vDash A$.

2. Use the Compactness Theorem to prove that a partial ordering R on a set S can be extended to a total ordering on that set. (*Hint*: Let $\mathscr{V}$ be the vocabulary consisting of the 2-place predicate symbol $<$ and constant symbols for each member of the set S. Let X be the set of formulae of $\mathscr{L}(\mathscr{V})$ consisting of the axioms for a total ordering (cf. page 86, S1–S3) and all formulae of the form $c_\alpha < c_\beta$ where c_α and c_β are constant symbols which correspond to members of S which are in the relation R. Prove that every finite subset of X is satisfiable.)

4.1 Logical closures

We now use Theorem I.11 to prove a few general results about logical closures.

THEOREM I.14. *Let C be any logical closure on $\mathscr{L}$. For any $A \in F_\mathscr{L}$ and any $X \subseteq F_\mathscr{L}$, if $X \vDash A$ then $A \in C(X)$.*

Proof. Assume $X \vDash A$. Then the set $X \cup \{\neg A\}$ is unsatisfiable. Consequently by Theorem I.11, $f \in C(X, \neg A)$. By L1, this implies $A \in C(X)$.

COROLLARY. *For any $X \subseteq F_{\mathscr{L}}$ and any $A \in F_{\mathscr{L}}$, $A \in C_s(X)$ if and only if $X \vDash A$.*

Proof. If $A \in C_s(X)$, then $X \vDash A$ by the definition of C_s. If $X \vDash A$, then $A \in C_s(X)$ by Theorem I.14.

It is reasonable to ask the following question. Are there any logical closures for $\mathscr{L}$ which do not coincide with C_s, and if so, what are they?

The first half of this question can be answered easily. Suppose that X_0 is any fixed set of formulae in the language $\mathscr{L}$. If we define C by

(1) $$C(X) = C_s(X \cup X_0), \text{ for any } X \subseteq F_{\mathscr{L}},$$

then it is easy to prove that C is a logical closure. (The details of this proof are left as an exercise. The only slightly difficult part is verifying that C satisfies property C2.) If the formulae in X_0 are not all logically valid, then C does not coincide with C_s.

The interesting fact is that every logical closure for the language $\mathscr{L}$ has the form (1), as we shall prove in the following theorem.

THEOREM I.15. *If C is a logical closure for $\mathscr{L}$, then for every $X \subseteq F_{\mathscr{L}}$*

$$C(X) = C_s(X \cup C(\emptyset)).$$

We first prove two lemmas.

LEMMA 1. *If C is a logical closure for $\mathscr{L}$, and α is a conjunctive formula of $\mathscr{L}$, then for any $X \subseteq F_{\mathscr{L}}$,*

$$C(X, \alpha) = C(X, \alpha_1, \alpha_2).$$

Proof. Since $\alpha \in C(X, \alpha)$, then by L2, $\alpha_1, \alpha_2 \in C(X, \alpha)$. Also $X \subseteq C(X, \alpha)$. Hence $X \cup \{\alpha_1, \alpha_2\} \subseteq C(X, \alpha)$. Thus, by C2 and C3 $C(X, \alpha_1, \alpha_2) \subseteq C(X, \alpha)$. Conversely, since $\alpha_1, \alpha_2 \in C(X, \alpha_1, \alpha_2)$, then by L2, $\alpha \in C(X, \alpha_1, \alpha_2)$. Thus $X \cup \{\alpha\} \subseteq C(X, \alpha_1, \alpha_2)$, and $C(X, \alpha) \subseteq C(X, \alpha_1, \alpha_2)$.

LEMMA 2. *For any logical closure C for $\mathscr{L}$, any $X \subseteq F_{\mathscr{L}}$, and any $A, B \in F_{\mathscr{L}}$:*
(i)   *if $B \in C(X, A)$, then $A \to B \in C(X)$;*
(ii) *if $A \in X$ and $A \to B \in C(X)$, then $B \in C(X)$.*

Proof. (i): Since $B \in C(X, A)$, then $f \in C(X, A, \neg B)$, and by Lemma 1, $f \in C(X, \neg(A \to B))$. Hence $A \to B \in C(X)$.
(ii): Since $A \to B \in C(X)$, then $f \in C(X, \neg(A \to B))$, and by Lemma 1, $f \in C(X, A, \neg B)$. But since $A \in X$, then $f \in C(X, \neg B)$, and it follows that $B \in C(X)$.

To prove Theorem I.15, we want to show that for any $X \subseteq F_{\mathscr{L}}$,

$$C(X) = C_s(X \cup C(\emptyset)).$$

First take any $A \in C(X)$. By C4, there exists a finite subset Y of X such that $A \in C(Y)$. If $A_1, A_2, \ldots, A_n$ are the members of Y, $A \in C(A_1, A_2, \ldots, A_n)$, and by n applications of Lemma 2(i),

$$A_1 \to \ldots \to A_n \to A \in C(\emptyset).$$

By Lemma 2(ii)

$$A_2 \to \ldots \to A_n \to A \in C_s(X \cup C(\emptyset)),$$

and by repeated applications of this lemma, we obtain $A \in C_s(X \cup C(\emptyset))$. Conversely, take any $A \in C_s(X \cup C(\emptyset))$. Thus $X \cup C(\emptyset) \vDash A$ and by Theorem I.14 $A \in C(X \cup C(\emptyset))$. But $X \subseteq C(X)$ and $C(\emptyset) \subseteq C(X)$, so that it follows that $C(X \cup C(\emptyset)) \subseteq C(X)$. Therefore, $A \in C(X)$. This completes the proof of the theorem.

COROLLARY. *Let C be any function from the power set of $F_{\mathscr{L}}$ into itself. Then C is a logical closure for $\mathscr{L}$ if and only if for every $X \subseteq F_{\mathscr{L}}$*

$$C(X) = C_s(X \cup C(\emptyset)).$$

5 Alternative interpretations of the ε-symbol

In §3.3 we formulated the semantics of a language $\mathscr{L}$ by interpreting the ε-symbol as a choice function Φ for a universe M, such that Φ assigns to the empty set some arbitrary but fixed member of M. We shall now discuss the general problem of finding a suitable interpretation for the ε-symbol and the particular solutions to this problem which appear in the literature.

Since Hilbert introduced the ε-symbol merely as a formal syntactic device to facilitate proof-theoretic investigations of the predicate calculus and of mathematical theories, such as arithmetic, which are based on the predicate calculus, the status of the ε-symbol is somewhat different from that of the other logical primitives. Although the basic methodology of Hilbert's formalist programme is to treat all symbols as meaningless, there is little doubt as to the *intended* interpretation of the symbols $^-$, $\to$, &, $\vee$, $\sim$, E, and ( ). On the other hand, it is by no means clear what interpretation is intended for the ε-symbol, or whether, in fact, any interpretation is intended. Hilbert's informal remarks about this symbol amount to little more than 'εxA is some object of the domain of individuals, such that if anything satisfies the formula A, then εxA does' ([1939], page 12). Hilbert's main concern is with the formal system which is obtained when the ε-symbol and the ε-formula,

$$A(a) \to A(\varepsilon x A),$$

are adjoined to the predicate calculus, and the rule of substitution is extended to allow for the replacement of free variables by ε-terms. The significance of his ε-Theorems is that a given deduction in the predicate calculus can be converted, using the ε-calculus, into another deduction of a certain special form in the predicate calculus. Consequently, the question of an interpretation, or even an intended interpretation, of the ε-symbol is unimportant both to his methods and to his results. The intended interpretation of the other logical symbols is unimportant only with respect to his methods.

Although Hilbert provides us with no precise semantic interpretation of the ε-symbol, his formal system does suggest certain properties which any interpretation must satisfy. First of all, the ε-formula and its deductive equivalent

$$(\varepsilon_0) \qquad\qquad \exists x A \rightarrow A(\varepsilon x A)$$

must be valid under this interpretation. Secondly, the interpretation must assign a value to every term of the form $\varepsilon x A$, even when the formula A is not satisfiable. This second requirement follows from the fact that Hilbert's system includes a rule for the replacement of free variables by arbitrary ε-terms. A third requirement would be that the 'second ε-axiom schema'

$$(\varepsilon_2) \qquad\qquad \forall x(A \leftrightarrow B) \rightarrow (\varepsilon x A = \varepsilon x B)$$

should be valid under this interpretation. Although formulae of this form are not taken as axioms in Hilbert's system, schema (ε_2) is a standard axiom schema in formalizations of set theory which incorporate the ε-symbol. We shall say that an interpretation of the ε-symbol is 'suitable' if it satisfies these three conditions.

The idea of using choice functions as an interpretation of this symbol was first investigated by Asser [1957] following a suggestion by Schröter. His investigations deal with three types of choice function. The first type is the one which we have used in §3.3. This interpretation is suitable in view of our results in that section.

Even if one agrees to interpret the ε-symbol in terms of a choice function, there are various ways of interpreting a 'null term'—that is, an ε-term $\varepsilon x A$, where A is unsatisfiable. Clearly, the interpretation of such a term depends on the entity, if any, which the choice function assigns to the null set. The choice function we have used (Asser's first type) assigns to the null set some arbitrary, but fixed member of the universe M. Two alternative definitions have been proposed: (i) Asser's second type of choice function, where Φ is undefined on the null set; (ii) the choice function employed by Hermes [1965] where Φ assigns the same value to the null set as it does to the universe, i.e.

$$(1) \qquad\qquad \Phi(\emptyset) = \Phi(M).$$

Hermes' definition seems to have been prompted by the way Hilbert and Bernays initially define the ε-symbol in terms of the η-symbol. The η-symbol is formally introduced by means of the following η-*rule* ([1939], page 10).

'If a formula $\exists xA$ is an axiom or is derivable, then ηxA can be introduced as a term, and the formula $A(\eta xA)$ can be taken as an initial formula', i.e. from $\exists xA$, one can infer $A(\eta xA)$.

Obviously, the η-symbol represents the 'indefinite article' in the same way that Russell's ι-symbol represents the 'definite article'. The ε-symbol is then defined as follows (p. 11):

(2) $$\varepsilon xA =_{\text{Df}} \eta x(\exists y[A]_y^x \rightarrow A).$$

From this definition it follows by the η-rule and the predicate calculus that any formula of the form

$$\exists xA \rightarrow A(\varepsilon xA)$$

is derivable. Furthermore, although ηxA is a term only if $\exists xA$ is derivable, any expression of the form εxA is a term. Hilbert and Bernays then dispense with the η-symbol, and instead take the ε-symbol as a primitive and introduce the ε-formula as an axiom.

Although this method of introducing the ε-symbol is only a heuristic device, Asser (p. 65) and presumably Hermes see in it an indication of Hilbert's intended interpretation of a null term. For, suppose there is no x for which A holds, then the formula $\exists y[A]_y^x \rightarrow A$ is true for all x. Consequently, using a choice function interpretation for both η and ε, it would follow from definition (2) that this choice function must assign the same value to the null set as it does to the universe.

If we define a model using Hermes' notion of a choice function, our results still hold subject to the following modifications.

In the definition of a logical closure the following additional condition is required:

L7. $(\varepsilon x(x = x) = \varepsilon x \neg(x = x)) \in C(X).$

In the proof of theorem I.11 it is necessary to show that the function Φ as defined on page 27 satisfies Hermes' condition

$$\Phi(\emptyset) = \Phi(M).$$

This follows immediately from L7 and the fact that $\langle (x = x), x \rangle$ represents M and $\langle \neg(x = x), x \rangle$ represents $\emptyset$. It is not difficult to show that C_s (under our new definition of a model) satisfies L7. Hence the Compactness Theorem still holds.

The second type of choice function which Asser considers is that which is undefined on the null set. It is clear from his results that this concept of choice function is better suited as an interpretation of the η-symbol than of the ε-symbol, since no value is assigned to a null term and thus our second requirement for a 'suitable' interpretation fails.

Asser points out that the ε-calculus of Hilbert and Bernays is not complete under his first interpretation of the ε-symbol. In an attempt to find an interpretation under which their ε-calculus is complete, Asser defines a third type of choice function, which is a very complicated modification of the first.

'Wir werden nun zeigen, daß es tätsachlich möglich ist, den Begriff der Auswahlfunktion so zu fassen, daß die zugehörige Interpretation dem formalen Ansatz von Hilbert adäquat ist. Allerdings ist dieser Begriff von Auswahlfunktion so kompliziert, daß sich seine Verwendung in der inhaltlichen Mathematik kaum empfiehlt.'[1]

Later (p. 65) Asser remarks that in view of the complexity of this third type of choice function, it is unlikely that this was Hilbert's intended interpretation of the ε-symbol.

We have chosen to interpret the ε-symbol in terms of Asser's first type of choice function for three reasons: (i) this interpretation is intuitively natural and simple to define (as opposed to Asser's third interpretation); (ii) this interpretation satisfies the three requirements we have given for a 'suitable' interpretation; (iii) under this interpretation the ε-calculus which is used in formalizing set theory (cf. Ackermann [1937–8] and Bourbaki [1954]) is complete. Although Hermes' interpretation is also 'suitable', we have not used it because it then becomes necessary to adjoin a new axiom, such as

$$\varepsilon x(x = x) = \varepsilon x \neg (x = x),$$

to the axioms of the ε-calculus in order to maintain completeness. There does not, in general, seem to be any good reason why such an axiom should be available in the ε-calculus.

However, it would be a mistake to state dogmatically that one particular interpretation of the ε-symbol is correct and all others are incorrect. One of the advantages of this symbol is its flexibility and indeterminacy. It is always possible, and often advantageous, to adjoin additional ε-axioms to a system, thereby making the designations of the ε-terms more definite. For example, Hilbert and Bernays have shown (pages 85–87) that by taking the formula (schema)

(1) $A(t) \to \varepsilon x A \neq t'$

---

[1] Asser (1957), p. 59.

as an axiom schema of arithmetic, the principle of mathematical induction can be derived. (The symbol $'$ denotes the arithmetic successor function.) In this formalization of arithmetic, the ε-symbol can be interpreted as a least number operator, although this is not the only possible interpretation, since (1) still allows the ε-symbol a certain amount of indeterminacy. If (1) is replaced by the stronger axiom schema

$$(2) \qquad\qquad A(t) \rightarrow \varepsilon x A \leqslant t,$$

then the ε-symbol is uniquely characterized as a least number operator (cf. Tait [1965]). We shall return to these applications of the ε-symbol to arithmetic in Chapter IV, §3.2.

6 Languages without identity

Although this book deals only with languages whose logical constants include the identity symbol $=$, it is easy to modify the definitions and proofs so that our main results also hold for languages without an identity symbol.

In the definition of a logical closure for a language without identity we replace conditions L5 and L6 by the following:

L8. If $\forall z(A_1 {}_z^{x_1} \leftrightarrow A_2 {}_z^{x_2}) \in X$ and $B^y_{\varepsilon x_1 A_1} \in X$, then $B^y_{\varepsilon x_2 A_2} \in C(X)$, where B is any atom and y any variable which does not have a free occurrence in B within the scope of an ε-symbol.

Theorem I.11 still holds for languages without identity. In the proof a more complicated definition of the equivalence relation $\sim$ is required. Observe that for any term s, one of the following conditions must hold: (i) s is an individual symbol, (ii) s is an ε-term, (iii) s is of the form $gs_1 \ldots s_n$, where $n \geqslant 0$. The definition of $s \sim t$ is by induction on the length of s as follows.

Case 1. s is an individual symbol a: Then $s \sim t$ iff t is the same individual symbol.

Case 2. s is of the form $\varepsilon x A$: Then $s \sim t$ iff t is of the form $\varepsilon y B$ and for all $r \in T_{\mathscr{L}}$, $A_r^x \in X$ iff $B_r^y \in X$.

Case 3. s is of the form $gs_1 \ldots s_n$: Then $s \sim t$ iff t is of the form $gt_1 \ldots t_n$ and $s_i \sim t_i$, for each $i = 1, \ldots, n$.

Lemma 1, that $\sim$ is an equivalence relation, follows easily from the definition of $\sim$. Lemma 2 follows by L8, and Lemma 3 is immediate from the definitions. The rest of the proof of Theorem I.11 goes through unchanged.

CHAPTER II

FORMAL SYSTEMS

1.1 Finitary reasoning

In Chapter I we formalized the intuitive notion of logical consequence by using the non-constructive techniques of set theory to define the semantic consequence relation ⊨. Although this approach has a certain abstract mathematical appeal, it is important, particularly in proving the consistency of mathematical theories, to find a more concrete definition of logical consequence. For this reason we now turn to the notion of deducibility in a formal system.

A formal system may be regarded as an array of uninterpreted symbols together with rules for manipulating these symbols. Consequently, in dealing with formal systems we can use a much weaker metatheory than that which was used in dealing with models. Throughout the present chapter and succeeding chapters nearly all our metatheoretic arguments will fall within the domain of what Hilbert calls *finitary reasoning (das finite Schließen)*. Hilbert defines this type of reasoning as follows: (Hilbert and Bernays [1934], page 32, translation by Kneebone [1963] page 205):

'We shall always use the word "finitary" to indicate that the discussion, assertion, or definition in question is kept within the bounds of thorough-going producibility of objects and thorough-going practicability of processes, and may accordingly be carried out within the domain of concrete inspection.'

In other words our discussions will deal with concrete objects such as terms, formulae, and finite sequences of formulae. In order to prove that a certain concrete object exists we must exhibit that object or at least describe a procedure for finding or constructing such an object. For example a proof of a metalinguistic statement of the form 'for all x, there exists a y such that . . . ' is a finitary proof if it enables one to construct an appropriate y for any given x.

Throughout this chapter we shall use the basic facts about the natural numbers and in particular the principle of mathematical induction. When we do so our metatheory remains finitary, since a natural number may be regarded as a sequence of vertical strokes and if we have proved by induction that every natural number has a certain property, then for any given number n, the inductive proof provides a method for showing in a finite number of steps that n has that property. (See Hilbert and Bernays [1934].)

We shall continue to use the *language* of set theory as part of our meta-language. Thus, for example, we will speak of 'sets' of formulae. This use of set theoretic terminology is only a matter of convenience, and nearly all references to sets could be eliminated.

The only theorems of this chapter which are not proved by finitary techniques are Theorems II.1 and II.11 which deal with the completeness of formal systems. In these two cases a finitary proof is impossible since the very notion of completeness depends on the non-finitary notion of semantic consequence. Similarly, Exercise 3 at the end of §2, which deals with the soundness of the ε-calculus, cannot be proved by finitary reasoning.

1.2 General definitions

In general, we may say that a *formal system* $\mathscr{F}$ for a vocabulary $\mathscr{V}$ consists of certain concrete objects called *deductions*, or more precisely *deductions of A from X*, where A is a formula of $\mathscr{L}(\mathscr{V})$ and X is a set of formulae of $\mathscr{L}(\mathscr{V})$. For the formal systems which we are about to consider, the deductions of A from X are certain finite sequences of formulae whose last member is A. However, in Chapter V we deal with a formal system whose deductions are sequences of sequences of formulae. Although particular formal systems can be set up in a variety of different ways, the one important feature which is shared by every formal system is the existence of an effective procedure for determining whether or not a given array of formulae is a deduction of A from X in that system.

If there exists a deduction of A from X in $\mathscr{F}$, then A is said to be *deducible from X in $\mathscr{F}$* and we denote this by writing $X \vdash_{\mathscr{F}} A$. Thus the statement expressed by the notation $X \vdash_{\mathscr{F}} A$ is an existential statement in the metalanguage, and a finitary proof of such a statement must provide a procedure for constructing a deduction of A from X. A deduction of A from $\emptyset$ in $\mathscr{F}$ is called a *proof of A in $\mathscr{F}$*. If there exists a proof of A in $\mathscr{F}$, then A is said to be a *theorem* of $\mathscr{F}$. A formal system $\mathscr{F}$ for $\mathscr{V}$ is *consistent* if f is not a theorem of $\mathscr{F}$, *sound* if $X \vdash_{\mathscr{F}} A$ implies $X \vDash A$, and *complete* if $X \vDash A$ implies $X \vdash_{\mathscr{F}} A$, for every $X \subseteq F_{\mathscr{L}(\mathscr{V})}$ and every $A \in F_{\mathscr{L}(\mathscr{V})}$. (This definition of completeness applies only to formal systems which incorporate the ε-symbol. For a formal system $\mathscr{F}$, such as the predicate calculus, where the ε-symbol may not be used in a deduction, we say that $\mathscr{F}$ is *complete* if $X \vDash A$ implies $X \vdash_{\mathscr{F}} A$ for every ε-free formula A of $\mathscr{L}(\mathscr{V})$ and every set X of ε-free formulae of $\mathscr{L}(\mathscr{V})$.)

If $\mathscr{F}$ is a formal system for $\mathscr{V}$, we define the *deductive closure C* for $\mathscr{F}$ as follows. For any set X of formulae in $\mathscr{L}(\mathscr{V})$

$$C(X) = \{A : X \vdash_{\mathscr{F}} A\}.$$

The following theorem, which depends on the Satisfiability Theorem, provides a useful method for proving the completeness of a formal system.

THEOREM II.1. *For any vocabulary $\mathscr{V}$ and any formal system $\mathscr{F}$ for $\mathscr{V}$, if the deductive closure C for $\mathscr{F}$ is a logical closure for $\mathscr{L}(\mathscr{V})$, then $\mathscr{F}$ is complete.*

Proof (non-finitary). Suppose $X \vDash A$, where $X \subseteq F_{\mathscr{L}(\mathscr{V})}$ and $A \in F_{\mathscr{L}(\mathscr{V})}$. Then by Theorem I.14, $A \in C(X)$. Hence $X \vdash_{\mathscr{F}} A$, and $\mathscr{F}$ is complete.

1.3   Axioms and rules of inference

Let $\mathscr{V}$ be any vocabulary. A particular formal system $\mathscr{F}$ for $\mathscr{V}$ is often defined by specifying certain formulae of $\mathscr{L}(\mathscr{V})$ as *axioms* and by prescribing certain *rules of inference,* i.e., rules which determine effectively whether a given formula of $\mathscr{L}(\mathscr{V})$ 'follows from' other given formulae of $\mathscr{L}(\mathscr{V})$. The formal concept of a deduction in $\mathscr{F}$ is defined in the obvious way. For any set X of formulae of $\mathscr{L}(\mathscr{V})$ and any formula A of $\mathscr{L}(\mathscr{V})$, a *deduction* of A from X in $\mathscr{F}$ is a sequence $\langle A_1, \ldots, A_n \rangle$ of formulae of $\mathscr{L}(\mathscr{V})$ such that A_n is A and for each $i = 1, \ldots, n$, A_i is an axiom, or A_i is a member of X, or A_i follows by some rule of inference from some preceding members of the sequence.

In order that $\mathscr{F}$ be effectively defined it is necessary to give an effective definition of the axioms of $\mathscr{F}$. Since in most cases we wish to specify infinitely many formulae as axioms, it is impossible to list all the axioms. However, we can specify a finite number of *forms* and then say that every axiom must be of one of these forms. The forms themselves are called *axiom schemata.* For example, we may say that any formula which has the form

$$A \to B \to A$$

is an axiom. Then the metalinguistic expression '$A \to B \to A$' is an axiom schema. Any formula which has this form is called an *instance* of the axiom schema.

We shall now use the above approach to define a formal system called the *ε-calculus for $\mathscr{V}$.*

2   The ε-calculus for $\mathscr{V}$

Let $\mathscr{V}$ be any vocabulary. The *ε-calculus for $\mathscr{V}$*, which we shall denote by $\varepsilon(\mathscr{V})$, is defined as follows.

The *axioms* of $\varepsilon(\mathscr{V})$ are all formulae of $\mathscr{L}(\mathscr{V})$ which are instances of the following axiom schemata:

P1 $A \to B \to A$
P2 $(A \to B \to C) \to (A \to B) \to A \to C$
P3 $(\neg A \to \neg B) \to (B \to A)$
P4 $(A \to f) \to \neg A$

P5 $(A \land B) \to A$

P6 $(A \land B) \to B$

P7 $A \to B \to (A \land B)$

P8 $\neg(A \lor B) \to \neg A$

P9 $\neg(A \lor B) \to \neg B$

P10 $\neg A \to \neg B \to \neg(A \lor B)$

Q1 $\forall x A \to \neg \exists x \neg A$

Q2 $\neg \forall x A \to \exists x \neg A$

Q3 $\neg \exists x A \to \neg A(t)$

Q4 $\exists x A \to A(\varepsilon x A)$

E1 $(s = t \land A_s^x) \to A_t^x$

E2 $\forall z(A_z^x \leftrightarrow B_z^y) \to \varepsilon x A = \varepsilon y B$

E3 $t = t$

Restriction: In axiom schema E1, A is an atom and x any variable which does not have a free occurrence in A within the scope of an ε-symbol.

In particular, any instance of axiom schemata P1–P10 is a *propositional axiom*, any instance of Q1–Q4 is a *quantificational axiom*, and any instance of E1–E3 is an *equality axiom*. Furthermore, an instance of P1 is called a P1-*axiom*, an instance of P2 a P2-*axiom*, etc.

There is one rule of inference for $\varepsilon(\mathscr{V})$. A formula B *follows by modus ponens from A and C* if and only if C is of the form $A \to B$. This rule can be expressed schematically as follows.

$$modus \ ponens: \quad \frac{A, \ A \to B}{B}$$

The deductions in the ε-calculus for $\mathscr{V}$ are now defined in the usual way. For any set X of formulae of $\mathscr{L}(\mathscr{V})$ and any formula A of $\mathscr{L}(\mathscr{V})$ a *deduction* $\mathscr{D}$ of A from X in $\varepsilon(\mathscr{V})$ is any finite sequence $\langle A_1, \ldots, A_n \rangle$ of formulae of $\mathscr{L}(\mathscr{V})$ such that A_n is A and for each $i = 1, \ldots, n$ at least one of the following conditions holds:

(i) A_i is an axiom of $\varepsilon(\mathscr{V})$,

(ii) A_i is a member of X,

(iii) A_i follows by modus ponens from A_j and A_k, for some $j, k < i$.

If A_i is a member of X, then A_i is called an *assumption* of $\mathscr{D}$, and if A_i satisfies condition (i) but not conditions (ii) or (iii), then A_i is said to be *used as an axiom* in $\mathscr{D}$. In other words a formula in $\mathscr{D}$ is used as an axiom if its presence in $\mathscr{D}$ can be justified only by (i).

If there exists a deduction of A from X in $\varepsilon(\mathscr{V})$, we say that A is *deducible* from X in $\varepsilon(\mathscr{V})$ and denote this fact by writing $X \vdash_{\varepsilon(\mathscr{V})} A$. To simplify the notation we write $X, B_1, \ldots, B_n \vdash_{\varepsilon(\mathscr{V})} A$ instead of $X \cup \{B_1, \ldots, B_n\} \vdash_{\varepsilon(\mathscr{V})} A$ and $\vdash_{\varepsilon(\mathscr{V})} A$ instead of $\emptyset \vdash_{\varepsilon(\mathscr{V})} A$.

Since $\mathscr{V}$ is an arbitrary vocabulary, the above definition of the formal system $\varepsilon(\mathscr{V})$ determines a whole class of formal systems, one for each vocabulary $\mathscr{V}$. Throughout this chapter we shall let $\mathscr{V}$ be a fixed, but arbitrary vocabulary, and we shall often speak of 'the ε-calculus' instead of 'the ε-calculus for $\mathscr{V}$' and write $X \vdash_\varepsilon A$, instead of $X \vdash_{\varepsilon(\mathscr{V})} A$. Furthermore, when it is understood that we are referring to deductions in the ε-calculus (for $\mathscr{V}$) we shall write simply $X \vdash A$.

Recall that a *proof* in a formal system $\mathscr{F}$ is a deduction from the empty set, and that A is a *theorem* of $\mathscr{F}$ if there exists a proof of A. We now give an example of a proof in $\varepsilon(\mathscr{V})$.

THEOREM II.2. *For any formula A of $\mathscr{L}(\mathscr{V})$, the formula $A \to A$ is a theorem of $\varepsilon(\mathscr{V})$.*

Proof. The following sequence of formulae constitutes a proof of $A \to A$.

(1) $A \to (A \to A) \to A$ P1-axiom
(2) $(A \to (A \to A) \to A) \to (A \to A \to A) \to A \to A$ P2-axiom
(3) $(A \to A \to A) \to A \to A$ modus ponens from (1) and (2)
(4) $A \to A \to A$ P1-axiom
(5) $A \to A$ modus ponens from (4) and (3)

THEOREM II.3. *Let X and Y be any sets of formulae of $\mathscr{L}(\mathscr{V})$ and $A, B_1, \ldots,$ B_n any formulae of $\mathscr{L}(\mathscr{V})$. Then:*
  (i)   *If A is an axiom of $\varepsilon(\mathscr{V})$ or a member of X, then $X \vdash A$.*
  (ii) *If $Y, B_1, \ldots, B_n \vdash A$, and $X \vdash B_i$ for each $i = 1, \ldots, n$, then $X \cup Y \vdash A$.*
  (iii) *If $X \subseteq Y$ and $X \vdash A$, then $Y \vdash A$.*
  (iv) *If $X \vdash A$, then there exists a finite subset X' of X such that $X' \vdash A$.*

Proof. Parts (i), (iii), and (iv) follow immediately from the definition of a deduction. To prove part (ii) it is sufficient to consider the case where $n = 1$, since the general case then follows by induction. Let $\langle A_1, \ldots, A_m \rangle$ be a deduction of B_1 from X and let $\langle C_1, \ldots, C_n \rangle$ be a deduction of A from $Y \cup \{B_1\}$. Then the sequence $\langle A_1, \ldots, A_m, C_1, \ldots, C_n \rangle$ is a deduction of A from $X \cup Y$.

EXERCISES

1. Using only Theorem II.3, prove that the deductive closure C for $\varepsilon(\mathscr{V})$ is a finitary closure operation on the set of formulae of $\mathscr{L}(\mathscr{V})$.

2. Prove that every propositional axiom of $\varepsilon(\mathscr{V})$ is a tautology.

3. Prove that $\varepsilon(\mathscr{V})$ is sound, i.e., for any set X of formulae of $\mathscr{L}(\mathscr{V})$ and any formula A of $\mathscr{L}(\mathscr{V})$ prove that $X \vdash_{\varepsilon(\mathscr{V})} A$ implies $X \vDash A$.

3   Derived rules of inference

In general, it is very impractical to prove that there exists a deduction of A from X by actually displaying the appropriate sequence of formulae (as we did in proving Theorem II.2). It is far more convenient to have at our disposal certain basic rules which assert that a deduction of A from X can be formed from certain known deductions. Such rules are called *derived rules of inference.* For example, Theorem II.3(ii) asserts a useful rule of this type. The following theorem provides two rather obvious derived rules of inference.

THEOREM II.4. *Let X and Y be any sets of formulae. Then*:
(i) $\rightarrow$-elimination rule: *If $X \vdash A \rightarrow B$, then $X, A \vdash B$.*
(ii) MP rule: *If $Y \vdash B_1 \rightarrow \ldots \rightarrow B_n \rightarrow A$, and $X \vdash B_i$ for each $i = 1, \ldots, n$, then $X \cup Y \vdash A$.*

Proof. (i) Let $\langle A_1, \ldots, A_n \rangle$ be a deduction of $A \rightarrow B$ from X. Then the sequence $\langle A_1, \ldots, A_n, A, B \rangle$ is a deduction of B from $X \cup \{A\}$, since B follows by modus ponens from A and A_n.
(ii) Since $Y \vdash B_1 \rightarrow \ldots \rightarrow B_n \rightarrow A$, then $Y, B_1, \ldots, B_n \vdash A$ by n applications of the $\rightarrow$-elimination rule. Hence $X \cup Y \vdash A$ by Theorem II.3(ii).

In the next section we shall prove the converse of part (i). Although this result is commonly called the Deduction Theorem, we shall refer to it as the $\rightarrow$-introduction rule in order to emphasize its role as a derived rule of inference.

3.1   The $\rightarrow$-introduction rule (Deduction Theorem)

In mathematics one commonly proves a statement of the form 'if A, then B' by taking A as an assumption and deducing B from A. The following theorem can be regarded as a formal justification of this method.

THEOREM II.5 (The $\rightarrow$-introduction rule). *If $X, A \vdash B$, then $X \vdash A \rightarrow B$.*

Proof. Let $\langle A_1, \ldots, A_n \rangle$ be a deduction of B from $X \cup \{A\}$. We shall prove by induction that for each i, $X \vdash A \rightarrow A_i$, thus proving $X \vdash A \rightarrow B$, since A_n is B.

Case 1. A_i is an axiom or a member of X: Hence
$$X \vdash A_i \qquad \text{Theorem II.3(i)}$$
$$\vdash A_i \rightarrow A \rightarrow A_i \quad \text{P1-axiom}$$
$$X \vdash A \rightarrow A_i \qquad \text{MP}$$

Case 2. A_i is the formula A:
$$\vdash A \rightarrow A_i \quad \text{Theorem II.2}$$
$$X \vdash A \rightarrow A_i \quad \text{Theorem II.3(iii).}$$

Case 3. A_i follows from A_j and A_k by modus ponens, where $j, k < i$: Then A_k is the formula $A_j \rightarrow A_i$. Hence

$$X \vdash A \rightarrow A_j \rightarrow A_i \quad \text{induction hypothesis}$$
$$X \vdash A \rightarrow A_j \quad\quad \text{induction hypothesis}$$
$$\vdash (A \rightarrow A_j \rightarrow A_i) \rightarrow (A \rightarrow A_j) \rightarrow A \rightarrow A_i$$
$$\text{P2-axiom}$$
$$X \vdash A \rightarrow A_i \quad\quad \text{MP}$$

COROLLARY (The syllogism rule). *If* $X \vdash A_1 \rightarrow A_2$, $X \vdash A_2 \rightarrow A_3, \dots,$ *and* $X \vdash A_{n-1} \rightarrow A_n$, *then* $X \vdash A_1 \rightarrow A_n$.

Proof. It is sufficient to prove the rule for the case where $n = 3$, since the general case then follows by induction on n.
Assume $\vdash A_1 \rightarrow A_2$ and $X \vdash A_2 \rightarrow A_3$. Then

$$X, A_1 \vdash A_2 \quad\quad \text{→-elimination}$$
$$X, A_1 \vdash A_3 \quad\quad \text{MP}$$
$$X \vdash A_1 \rightarrow A_3 \quad \text{→-introduction}$$

3.2   Some theorems of the ε-calculus

Our derived rules of inference can now be used to show that formulae of certain standard forms are theorems of the ε-calculus.

THEOREM II.6. *For any formulae A and B*:

(i) $\vdash \neg A \rightarrow A \rightarrow B,$
(ii) $\vdash \neg\neg A \rightarrow A,$
(iii) $\vdash A \rightarrow \neg\neg A.$

Proof.

(i) $\vdash \neg A \rightarrow \neg B \rightarrow \neg A$ P1-axiom
    $\vdash (\neg B \rightarrow \neg A) \rightarrow (A \rightarrow B)$ P3-axiom
    $\vdash \neg A \rightarrow A \rightarrow B$ syllogism

(ii) $\vdash \neg\neg A \rightarrow \neg A \rightarrow \neg\neg\neg A$ part (i)
    $\vdash (\neg A \rightarrow \neg\neg\neg A) \rightarrow (\neg\neg A \rightarrow A)$ P3-axiom
    $\vdash \neg\neg A \rightarrow \neg\neg A \rightarrow A$ syllogism
$\neg\neg A \vdash A$ →-elimination (twice)
    $\vdash \neg\neg A \rightarrow A$ →-introduction

(iii) $\vdash \neg\neg\neg A \rightarrow \neg A$ part (ii)
    $\vdash (\neg\neg\neg A \rightarrow \neg A) \rightarrow (A \rightarrow \neg\neg A)$ P3-axiom
    $\vdash A \rightarrow \neg\neg A$ MP

THEOREM II.7 (Contrapositive rules).
(i)   *If* $X \vdash \neg A \rightarrow \neg B$, *then* $X \vdash B \rightarrow A$.
(ii)   *If* $X \vdash \neg A \rightarrow B$, *then* $X \vdash \neg B \rightarrow A$.
(iii)   *If* $X \vdash A \rightarrow \neg B$, *then* $X \vdash B \rightarrow \neg A$.
(iv)   *If* $X \vdash A \rightarrow B$, *then* $X \vdash \neg B \rightarrow \neg A$.

Proof. Use axiom schema P3, Theorem II.6(ii), (iii), the MP rule, and the syllogism rule.

THEOREM II.8. *For any conjunctive formula* α:
  (i) $\vdash \alpha \rightarrow \alpha_1$,
  (ii) $\vdash \alpha \rightarrow \alpha_2$,
  (iii) $\vdash \alpha_1 \rightarrow \alpha_2 \rightarrow \alpha$.

Proof. If α is of the form $A \wedge B$ or $\neg(A \vee B)$, then (i), (ii), and (iii) follow by the propositional axioms P5, P6, and P7, or P8, P9, and P10 respectively. Suppose α is of the form $\neg(A \rightarrow B)$.

(i) $\vdash \neg A \rightarrow A \rightarrow B$ Theorem II.6(i)
            $\vdash \neg(A \rightarrow B) \rightarrow A$ contrapositive

(ii) $\vdash B \rightarrow A \rightarrow B$ P1-axiom
            $\vdash \neg(A \rightarrow B) \rightarrow \neg B$ contrapositive

(iii) $A, A \rightarrow B \vdash B$ by modus ponens
            $A \vdash (A \rightarrow B) \rightarrow B$ $\rightarrow$-introduction
            $A \vdash \neg B \rightarrow \neg(A \rightarrow B)$ contrapositive
            $\vdash A \rightarrow \neg B \rightarrow \neg(A \rightarrow B)$ $\rightarrow$-introduction

THEOREM II.9 (The f-rules).
  (i) $X \vdash \neg A$ iff $X, A \vdash f$,
  (ii) $X \vdash A$ iff $X, \neg A \vdash f$.

Proof.
  (i) Assume $X \vdash \neg A$
            $\vdash \neg A \rightarrow A \rightarrow f$ Theorem II.6(i)
            $X \vdash A \rightarrow f$ MP
            $X, A \vdash f$ $\rightarrow$-elimination

      Assume $X, A \vdash f$
            $X \vdash A \rightarrow f$ $\rightarrow$-introduction
            $\vdash (A \rightarrow f) \rightarrow \neg A$ P4-axiom
            $X \vdash \neg A$ MP

(ii) $X \vdash A$ iff $X \vdash \neg\neg A$ Theorem II.6(ii), (iii) and MP
            iff $X, \neg A \vdash f$ part (i)

4 Completeness of the ε-calculus

So far everything which we have proved about the ε-calculus for $\mathscr{V}$ depends only on the facts that formulae of the form P1–P10 are axioms and that modus ponens is the one and only rule of inference. We shall now make use of the quantificational axioms.

THEOREM II.10. *For any universal formula γ, any term t, and any existential formula δ:*

(i) $\vdash \gamma \rightarrow \gamma(t)$,

(ii) $\vdash \delta \rightarrow \delta(\varepsilon\delta)$.

Proof. If γ is of the form $\neg\exists xA$, then $\gamma \rightarrow \gamma(t)$ is a Q3-axiom, and if δ is of the form $\exists xA$, then $\delta \rightarrow \delta(\varepsilon\delta)$ is a Q4-axiom. Suppose γ is $\forall xA$ and δ is $\neg\forall xA$.

(i) $\vdash \forall xA \rightarrow \neg\exists x\neg A$ Q1-axiom

$\vdash \neg\exists x\neg A \rightarrow \neg\neg A(t)$ Q3-axiom

$\vdash \neg\neg A(t) \rightarrow A(t)$ Theorem II.6(ii)

$\vdash \forall xA \rightarrow A(t)$ syllogism

(ii) $\vdash \neg\forall xA \rightarrow \exists x\neg A$ Q2-axiom

$\vdash \exists x\neg A \rightarrow \neg A(\varepsilon x\neg A)$ Q4-axiom

$\vdash \neg\forall xA \rightarrow \neg A(\varepsilon x\neg A)$ syllogism

THEOREM II.11 (The Completeness Theorem). *For any vocabulary $\mathscr{V}$ the ε-calculus for $\mathscr{V}$ is complete.*

Proof (non-constructive). By Theorem II.1, in order to prove that $\varepsilon(\mathscr{V})$ is complete it is sufficient to prove that its deductive closure C is a logical closure operation. That C is a finitary closure operation follows from Theorem II.3. The 'logical' properties L1, L2, L3, and L4 follow from Theorems II.9, II.8, II.10(i), and II.10(ii), respectively, using the MP rule. Properties L5 and L6 follow easily from the equality axioms E1 and E2.

Notice that our proof of the Completeness Theorem does not make use of the fact that formulae of the form

E3 $t = t$

are regarded as axioms of the ε-calculus. Although axioms of this form are superfluous in the ε-calculus, it is convenient to include them since they are needed in proving that the E2-axioms, i.e., formulae of the form

E2 $\forall z(A_z^x \leftrightarrow B_z^y) \rightarrow \varepsilon xA = \varepsilon yB$

can be eliminated from proofs of ε-free formulae.

EXERCISE

Prove constructively that if $X \vdash A$, then there exists a deduction of A from X in which no formula of the form $t = t$ is used as an axiom. (See the proof of Lemma 1(ix), page 26.)

5 The Tautology Theorem

Our proof of the Completeness Theorem is a good example of a non-constructive existence proof, since we prove that there exists a deduction of A

from X without actually describing how such a deduction can be formed. We now give a completely constructive proof of a much weaker completeness result—namely, that if A is a tautological consequence of $B_1, \ldots, B_n$, then there exists a deduction of A from $\{B_1, \ldots, B_n\}$.

We first establish some additional derived rules of inference.

THEOREM II.12. *Let X be any set of formulae, α any conjunctive formula, and β any disjunctive formula. Then*:

(i)   Contradiction rule: *if f is a member of X or if there exists some formula A such that both A and $\neg A$ are members of X, then $X \vdash f$.*

(ii) $\neg\neg$-rule: *if $X, A \vdash f$, then $X, \neg\neg A \vdash f$.*

(iii) α-rule: *if $X, \alpha_1, \alpha_2 \vdash f$, then $X, \alpha \vdash f$.*

(iv) β-rule: *if $X, \beta_1 \vdash f$ and $X, \beta_2 \vdash f$, then $X, \beta \vdash f$.*

Proof.

(i)   If f is a member of X, then $X \vdash f$ by Theorem II.3(i). Suppose A and $\neg A$ are members of X. Since $A \vdash A$, then $A, \neg A \vdash f$ by the f-rule. Hence $X \vdash f$ by Theorem II.3(iii).

(ii)   Assume $X, A \vdash f$
$$X \vdash \neg A \qquad f\text{-rule (i)}$$
$$X, \neg\neg A \vdash f \qquad f\text{-rule (ii)}$$

(iii)   Assume $X, \alpha_1, \alpha_2 \vdash f$
$$\alpha \vdash \alpha_1 \qquad \text{Theorem II.8(i) and } \rightarrow\text{-elimination}$$
$$\alpha \vdash \alpha_2 \qquad \text{Theorem II.8(ii) and } \rightarrow\text{-elimination}$$
$$X, \alpha \vdash f \qquad \text{Theorem II.3(ii)}$$

(iv)   Assume $X, \beta_1 \vdash f$ and $X, \beta_2 \vdash f$, where β_1 and β_2 are the disjunctive components of some disjunctive formula β. By the duality principle (page 16) the contrary of β is a conjunctive formula α such that α_1 and β_1 are contradictory and α_2 and β_2 are contradictory. Hence
$$X \vdash \alpha_1 \qquad\qquad f\text{-rule}$$
$$X \vdash \alpha_2 \qquad\qquad f\text{-rule}$$
$$\vdash \alpha_1 \rightarrow \alpha_2 \rightarrow \alpha \qquad \text{Theorem II.8(iii)}$$
$$X \vdash \alpha \qquad\qquad \text{MP}$$
$$X, \beta \vdash f \qquad\qquad f\text{-rule}$$

A finite set of formulae $\{A_1, \ldots, A_n\}$ is said to be *truth functionally invalid* if for every truth assignment ψ for $A_1 \wedge \ldots \wedge A_n$ there exists an A_i such that $\bar{\psi}(A_i) = 0$.

THEOREM II.13. *If the set $\{A_1, \ldots, A_n\}$ is truth functionally invalid, then $A_1, \ldots, A_n \vdash f$.*

Proof. Let m be the sum of the lengths of the A_i. The proof is by induction on m.

Case 1. Each of the A_i is either a molecule or the negation of a molecule: We shall prove that since $\{A_1, \ldots, A_n\}$ is truth functionally invalid, then either (i) some A_i is the formula f, or (ii) some A_i is the negation of some A_j. It then follows that $A_1, \ldots, A_n \vdash f$ by the contradiction rule. Suppose neither (i) nor (ii) holds. We can then define a truth assignment ψ for $A_1 \wedge \ldots \wedge A_n$ as follows. For each A_i, if A_i is a molecule let $\psi(A_i) = 1$, and if A_i is the negation of a molecule B_i let $\psi(B_i) = 0$. It then follows that for each i, $\bar{\psi}(A_i) = 1$, which contradicts the assumption that $\{A_1, \ldots, A_n\}$ is truth functionally invalid.

Case 2. At least one of the A_i is neither a molecule nor the negation of a molecule: We may assume that such a formula is A_1. Then one of the following three cases must hold.

Case 2a. A_1 is of the form $\neg\neg B$: In this case $\{B, A_2, \ldots, A_n\}$ must be truth functionally invalid since for any ψ, if $\bar{\psi}(\neg\neg B) = 0$, then $\bar{\psi}(B) = 0$. By the induction hypothesis $B, A_2, \ldots, A_n \vdash f$, and therefore $\neg\neg B, A_2, \ldots, A_n \vdash f$ by the $\neg\neg$-rule.

Case 2b. A_1 is a conjunctive formula α: In this case $\{\alpha_1, \alpha_2, A_2, \ldots, A_n\}$ must be truth functionally invalid since for any ψ, if $\bar{\psi}(\alpha) = 0$, then $\bar{\psi}(\alpha_1) = 0$ or $\bar{\psi}(\alpha_2) = 0$. By the induction hypothesis $\alpha_1, \alpha_2, A_2, \ldots, A_n \vdash f$, and therefore $\alpha, A_2, \ldots, A_n \vdash f$ by the α-rule.

Case 2c. A_1 is a disjunctive formula β: In this case both $\{\beta_1, A_2, \ldots, A_n\}$ and $\{\beta_2, A_2, \ldots, A_n\}$ are truth functionally invalid since for any ψ, if $\bar{\psi}(\beta) = 0$, then $\bar{\psi}(\beta_1) = 0$ and $\bar{\psi}(\beta_2) = 0$. Hence $\beta, A_2, \ldots, A_n \vdash f$ by the induction hypothesis and the β-rule.

Notice that the above theorem holds for any formal system which satisfies the four rules that make up Theorem II.12.

THEOREM II.14 (The Tautology Theorem). *Every tautology is a theorem of the ε-calculus.*

Proof. Let A be a tautology. Then $\{\neg A\}$ is truth functionally invalid and by Theorem II.13, $\neg A \vdash f$. Hence $\vdash A$ by the f-rule.

COROLLARY.
(i) *If A is a tautological consequence of $B_1, \ldots, B_n$, then $B_1, \ldots, B_n \vdash A$.*
(ii) *The tautology rule: if A is a tautological consequence of $B_1, \ldots, B_n$ and $X \vdash B_1, \ldots, X \vdash B_n$, then $X \vdash A$.*

Proof. Part (i) follows by the Tautology Theorem and the $\rightarrow$-elimination rule, and part (ii) by the Tautology Theorem and the MP rule.

6 The consistency of the ε-calculus

Recall that a formal system $\mathscr{F}$ is *consistent* if f is not a theorem of $\mathscr{F}$. Since f is not a valid formula, the soundness of the ε-calculus (cf. Exercise 3, page 41) implies its consistency. However this consistency proof is somewhat unsatisfactory since it involves the non-finitary notion of validity. For this reason, we shall now give a completely finitary proof of the consistency of the ε-calculus.

THEOREM II.15. *For any vocabulary $\mathscr{V}$, the ε-calculus for $\mathscr{V}$ is consistent.*

Proof. For any quasi-formula A of $\mathscr{L}(\mathscr{V})$, we define the formula $g(A)$ as follows by induction on the length of A.

(i)   If A is of the form $Pt_1, \ldots t_n$, then $g(A)$ is P, where P is now regarded as a 0-place predicate symbol.

(ii) If A is of the form $s = t$, then $g(A)$ is $\neg f$.

(iii) If A is the formula f, then $g(A)$ is f.

(iv) If A is of the form $\neg B$, or $B * C$, where $*$ is $\wedge$, $\vee$, or $\rightarrow$, then $g(A)$ is $\neg g(B)$, or $g(B) * g(C)$, respectively.

(v)   If A is of the form $\exists x B$ or $\forall x B$, then $g(A)$ is $g(B)$.

In other words, $g(A)$ is obtained from A by first erasing all the quasi-terms in A and all occurrences of the symbols $\forall x$ and $\exists x$, and then replacing each occurrence of $=$ by $\neg f$. It is easy to see that if A is an axiom of $\varepsilon(\mathscr{V})$, then $g(A)$ is a tautology, and if $g(A)$ and $g(A \rightarrow B)$ are tautologies, then $g(B)$ is a tautology. Consequently, if A is a theorem of the ε-calculus for $\mathscr{V}$, then $g(A)$ is a tautolology. Since $g(f)$ is not a tautology, then f is not a theorem, and therefore $\varepsilon(\mathscr{V})$ is consistent.

7.1   Some derived rules for operating with quantifiers

Theorem II.10 states that all formulae of the following forms are theorems of the ε-calculus: (i) $\forall x A \rightarrow A(t)$, (ii) $\neg \exists x A \rightarrow \neg A(t)$, (iii) $\neg \forall x A \rightarrow \neg A(\varepsilon x \neg A)$, and (iv) $\exists x A \rightarrow A(\varepsilon x A)$. These results provide the following derived rules of inference for the introduction and elimination of quantifiers.

THEOREM II.16. *Let X be any set of formulae. Then*:

(i) $\forall$-elimination rule: *if $X \vdash \forall x A$, then $X \vdash A(t)$ for any term t.*

(ii) $\exists$-introduction rule: *if there exists a term t such that $X \vdash A(t)$, then $X \vdash \exists x A$.*

(iii) $\forall$-introduction rule: *if $X \vdash A(\varepsilon x \neg A)$, then $X \vdash \forall x A$.*

(iv) $\exists$-elimination rule: *if $X \vdash \exists x A$, then $X \vdash A(\varepsilon x A)$.*

The proof follows immediately by Theorem II.10 and the tautology rule.

Often the specified ε-terms in the $\forall$-introduction rule and the $\exists$-elimination rule are rather complicated expressions. For example, if we know that $X \vdash \exists x \exists y Pxy$ and we then make two applications of the $\exists$-elimination rule we obtain the following complicated result

$$X \vdash P\varepsilon x \exists y Pxy \varepsilon y P\varepsilon x \exists y Pxyy.$$

In order to avoid such complicated expressions it is convenient to use metalinguistic symbols, such as the letters s and t, to denote certain specified ε-terms. For example, we can prove $\vdash \exists x \forall y Pxy \rightarrow \forall y \exists x Pxy$ as follows. Let X be the set $\{\exists x \forall y Pxy\}$. Then

$X \vdash \exists x \forall y Pxy$	Theorem II.3(i)
$X \vdash \forall y Psy$	$\exists$-elimination, s is $\varepsilon x \forall y Pxy$
$X \vdash Pst$	$\forall$-elimination, t is $\varepsilon y \neg \exists x Pxy$
$X \vdash \exists x Pxt$	$\exists$-introduction
$X \vdash \forall y \exists x Pxy$	$\forall$-introduction
$\vdash \exists x \forall y Pxy \rightarrow \forall y \exists x Pxy$	$\rightarrow$-introduction

EXERCISE

Prove $\vdash \exists x \exists y Pxy \rightarrow \exists y \exists x Pxy$ by using the letters s and t to denote the appropriate ε-terms.

An alternative way of avoiding complicated ε-terms is by employing the following rules of inference in which individual symbols are used in effect as abbreviations for arbitrary ε-terms.

THEOREM II.17. *Let X be any set of formulae, B any formula, A any quasi-formula which contains no free variable other than x, and a any individual symbol which does not appear in A, B, or any member of X. Then:*
(i)   *Substitution rule: if $X \vdash A_a^x$, then $X \vdash A_t^x$, for any term t.*
(ii) *Generalization rule: if $X \vdash A_a^x$, then $\vdash \forall x A$.*
(iii) *$\exists$-rule: if $\vdash A_a^x \rightarrow B$, then $X \vdash \exists x A \rightarrow B$.*
(iv) *$\forall$-rule: if $X \vdash B \rightarrow A_a^x$, then $X \vdash B \rightarrow \forall x A$.*

Proof.
 (i)   Let $\langle A_1, \ldots, A_n \rangle$ be a deduction of A_a^x from X. For each $i = 1, \ldots, n$, let A_i' be the formula obtained from A_i by replacing each occurrence of a in A_i by the term t. Since a does not occur in A and since A_n is A_a^x, then A_n' is A_t^x. It is easy to see that the sequence $\langle A_1', \ldots, A_n' \rangle$ constitutes a deduction of A_t^x from X. For, if A_i is an axiom, then A_i' is an axiom of the same form, if A_i is a member of X, then A_i' is A_i, and if A_i follows by modus ponens from A_j and A_k, then A_i' follows by that rule from A_j' and A_k'.
 (ii) Assume $X \vdash A_a^x$. Then $X \vdash A(\varepsilon x \neg A)$ by part (i). Hence $X \vdash \forall x A$ by the $\forall$-introduction rule.
 (iii) Assume $X \vdash A_a^x \rightarrow B$. Since a does not appear in B, the proof of part (i) yields $X \vdash A(\varepsilon x A) \rightarrow B$. Therefore $X \vdash \exists x A \rightarrow B$ by the syllogism rule and the Q4-axiom $\exists x A \rightarrow A(\varepsilon x A)$.
 (iv) The proof of (iv) is similar to that of (iii).

The proof of the above theorem depends on the crucial fact that when an individual symbol a is replaced by some term t in an axiom, the resulting formula is an axiom of the same form. The converse assertion is not in general true—that is, replacing a term t by an individual symbol a in an axiom does not necessarily yield another axiom of the same form. For example, the term t may be one of the specified ε-terms in a Q4-axiom or an E2-axiom, or it may contain the specified terms in a Q3- or Q4-axiom. The fact that the converse does not hold provides one of the major difficulties in proving the eliminability of the ε-symbol (Hilbert's Second ε-Theorem) as we shall see in the next chapter.

In view of these remarks we can now explain the motivation behind the restrictions which we imposed on the E1-axioms. Recall that a formula is an E1-axiom if it is of the form

$$(s = t \wedge A_s^x) \to A_t^x,$$

provided that A is an atom and *x is a variable which does not have a free occurrence in A within the scope of an ε-symbol*. This second restriction guarantees that if any ε-term is replaced by some other term t in an E1-axiom, the resulting formula is still an E1-axiom. Consequently in our proof of the Second ε-Theorem, the E1-axioms present no difficulties. (The reason for restricting A to an atom will be seen in Chapter V.)

In spite of these restrictions on the E1-axioms, the desired results concerning the identity symbol are deducible in the ε-calculus. In particular, we shall prove presently that *any* formula of the form

$$(s = t \wedge A_s^x) \to A_t^x$$

is a theorem of the ε-calculus.

We now use Theorem II.17 to establish another useful derived rule of inference.

THEOREM II.18 (The distribution rule). *Let X be any set of formulae, A and B any quasi-formulae, and a any individual symbol not appearing in A, B, or any member of X. If $X \vdash A_a^x \leftrightarrow B_a^y$, then*

(i)　$X \vdash \forall x A \leftrightarrow \forall y B,$

(ii)　$X \vdash \exists x A \leftrightarrow \exists y B,$

(iii)　$X \vdash \varepsilon x A = \varepsilon y B.$

Proof. (i) Assume $X \vdash A_a^x \leftrightarrow B_a^y$. Then

$$X \vdash A_a^x \to B_a^y \qquad \text{tautology rule}$$
$$X \vdash \forall x A \to A_a^x \qquad \text{Theorem II.10(i)}$$
$$X \vdash \forall x A \to B_a^y \qquad \text{syllogism rule}$$
$$X \vdash \forall x A \to \forall y B \qquad \forall\text{-rule.}$$

Similarly, $X \vdash \forall y B \to \forall x A$. Hence $X \vdash \forall x A \leftrightarrow \forall y B$ by the tautology rule.

(ii) The proof of (ii) is similar to that of (i).

(iii) Assume $X \vdash A_a^x \leftrightarrow B_a^y$. Let z be any variable which is free for x in A and free for y in B. Since $[A_z^x]_a^z$ is A_a^x and $[B_z^y]_a^z$ is B_a^y, then $X \vdash [A_z^x]_a^z \leftrightarrow [B_z^y]_a^z$. Hence $X \vdash \forall z(A_z^x \leftrightarrow B_z^y)$ by the generalization rule, and $X \vdash \varepsilon xA = \varepsilon yB$ by axiom schema E2 and the MP rule.

7.2   Substitution instances and universal closures

Let A be any quasi-formula and let $x_1, \ldots, x_n$ be the free variables in A. For any terms $t_1, \ldots, t_n$ the formula $A_{t_1 \cdots t_n}^{x_1 \cdots x_n}$ is called a *substitution instance* of A, and the formula $\forall x_1 \ldots \forall x_n A$ is called a *universal closure* of A. Thus A has $n!$ universal closures—one for each ordering of the variables in A. (If A has no free variables, then A is regarded as a substitution instance and a universal closure of itself.) For any X, we write $X \vdash \forall[A]$ to denote that every universal closure of A is deducible from X.

THEOREM II.19. *Let X be any set of formulae and A any quasi-formula. Then*:
  (i)   *If B is a universal closure of A and $X \vdash B$, then every substitution instance of A is deducible from X.*
  (ii) *If every substitution instance of A is deducible from X, then $X \vdash \forall[A]$.*

Proof. The proof of (i) follows by repeated application of the $\forall$-elimination rule and the proof of (ii) by repeated application of the $\forall$-introduction rule.

Theorem II.19(ii) has the following very useful application. Suppose we want to prove $\vdash \forall[A]$ for any quasi-formula A of a certain given form. If every substitution instance of A is also of this form, then by the above theorem it is sufficient to prove that every formula of this form is a theorem. In other words we may assume that A is a formula and simply prove $\vdash A$. We shall use this technique in proving the following theorem.

THEOREM II.20. *Let B and C be any quasi-formulae, x any variable, and y any variable which is free for x in B and does not occur free in either B or C. Let Q denote either $\forall$ or $\exists$. Then*:
  (i) $\vdash \forall[QxB \leftrightarrow Qy[B]_y^x]$ *and* $\vdash \forall[\varepsilon xB = \varepsilon y[B]_y^x]$;
  (ii) $\vdash \forall[\neg QxB \leftrightarrow Q'x \neg B]$, *where Q' is $\exists$ if Q is $\forall$ and Q' is $\forall$ if Q is $\exists$*;
  (iii) $\vdash \forall[(QxB \lor C) \leftrightarrow Qy(B_y^x \lor C)]$;
  (iv) $\vdash \forall[(C \lor QxB) \leftrightarrow Qy(C \lor B_y^x)]$.

Proof. In each of the four parts of this theorem we want to prove that if A is a quasi-formula of a certain form, then $\vdash \forall[A]$. Since in each case every substitution instance of A is a formula of the same form as that of A, then by the above remark we may assume that A is a formula and simply prove $\vdash A$.

(i) Let a be any individual symbol not appearing in B. Since y is free for x in B, but not free in B, then $[B_y^x]_a^y$ is B_a^x. Hence $\vdash B_a^x \leftrightarrow [B_y^x]_a^y$ by the Tautology Theorem. The desired results now follow by the distribution rule.

(ii) If Q is $\forall$, then $\neg QxB \leftrightarrow Q'x\neg B$ is a tautological consequence of the Q-axioms $\forall xB \rightarrow \neg \exists x \neg B$ and $\neg \forall xB \rightarrow \exists x \neg B$. Suppose Q is $\exists$. Let a be any individual symbol not appearing in B. Then

$$\vdash B_a^x \leftrightarrow \neg \neg B_a^x \qquad \text{Tautology Theorem}$$
$$\vdash \exists xB \leftrightarrow \exists x \neg \neg B \qquad \text{distribution rule}$$
$$\vdash \neg \forall x \neg B \leftrightarrow \exists x \neg \neg B \qquad \text{part (ii), where } Q \text{ is } \forall$$
$$\vdash \neg \exists xB \leftrightarrow \forall x \neg B \qquad \text{tautology rule}$$

(iii)   We shall give a proof for the case where Q is $\forall$. The other case can be proved similarly. The proof hinges on the fact that for any term t, $[B_y^x \vee C]_t^y$ is $B_t^x \vee C$. Let t be the term $\varepsilon y \neg (B_y^x \vee C)$ and s the term $\varepsilon x \neg B$.

$$\left.\begin{array}{l} \vdash \forall y(B_y^x \vee C) \rightarrow B_s^x \vee C \\ \vdash \forall xB \rightarrow B_t^x \end{array}\right\} \quad \text{Theorem II.10(i)}$$

$$\left.\begin{array}{l} \vdash B_s^x \rightarrow \forall xB \\ \vdash B_t^x \vee C \rightarrow \forall y(B_y^x \vee C) \end{array}\right\} \quad \begin{array}{l}\text{Theorem II.10(ii) and} \\ \text{the contrapositive rule}\end{array}$$

$$\vdash (\forall xB \vee C) \leftrightarrow \forall y(B_y^x \vee C) \qquad \text{tautology rule}$$

(iv)   The proof is similar to that of (iii).

7.3   The equivalence rule

In this section we establish a derived rule of inference, the equivalence rule, which asserts in effect that within a given formula any expression may be replaced by an equivalent expression. Before giving a precise statement of this rule we first introduce the following unifying notation. Suppose E_1 and E_2 are any two quasi-terms or any two quasi-formulae. The notation $E_1 \equiv E_2$ is used to denote the quasi-formula $E_1 = E_2$ if E_1 and E_2 are both quasi-terms, and the quasi-formula $E_1 \leftrightarrow E_2$, if they are both quasi-formulae.

THEOREM II.21 (The equivalence rule). *Let X be any set of formulae, E and E' any two quasi-terms or any two quasi-formulae, and A any formula or term containing some specified occurrence of E. Let A' be the expression obtained from A by replacing this specified occurrence of E by E'. If A' is a term or formula and if $X \vdash \forall [E \equiv E']$, then $X \vdash A \equiv A'$.*

Proof. The proof is by induction on the length of A. (We may assume without loss of generality that X is a finite set.)
Case 1. The specified occurrence of E in A is A itself: Then A' is E' and $X \vdash A \equiv A'$ by the hypothesis.

We now assume that Case 1 does not apply. Consequently one of the following cases must apply.
Case 2. A is of the form $Ps_1 \ldots s_n$ where P is an n-place function or predicate symbol: Then the specified occurrence of E in A must be contained within some s_i. Hence A' is of the form $Ps_1 \ldots s_{i-1}s_i's_{i+1} \ldots s_n$, and $X \vdash s_i = s_i'$ by the induction hypothesis. Let B be the quasi-formula

$$Ps_1 \ldots s_n \equiv Ps_1 \ldots s_{i-1} x s_{i+1} \ldots s_n.$$

Then $B_{s_i}^x$ is $A \equiv A$ and $B_{s_i'}^x$ is $A \equiv A'$. We prove $X \vdash B_{s_i'}^x$ as follows:

$X \vdash s_i = s_i'$	induction hypothesis
$\vdash B_{s_i}^x$	E3-axiom or Tautology Theorem
$X \vdash (s_i = s_i' \wedge B_{s_i}^x) \to B_{s_i'}^x$	E1-axiom
$X \vdash B_{s_i'}^x$	tautology rule

Case 3. A is of the form $s = t$: The proof is similar to that of Case 2.

Case 4. A is of the form $\neg B$: Then A' is of the form $\neg B'$, and $X \vdash B \leftrightarrow B'$ by the induction hypothesis. Hence $X \vdash \neg B \leftrightarrow \neg B'$ by the tautology rule.

Case 5. A is of the form $B \wedge C$, $B \vee C$, or $B \to C$: The proof is similar to that of Case 4.

Case 6. A is of the form $\exists x B$, $\forall x B$, or $\varepsilon x B$. Then A' is of the form $\exists x B'$, $\forall x B'$, or $\varepsilon x B'$, respectively, where B' is obtained from B by replacing some occurrence of E in B by E'. Let a be some individual symbol not appearing in B, B', or any member of X. It is sufficient to prove $X \vdash B_a^x \leftrightarrow B'{}_a^x$ since the desired result then follows by the distribution rule. If no free occurrence of x in B lies within the specified occurrence of E, then B_a^x contains this occurrence of E and $B'{}_a^x$ is obtained from B_a^x by replacing this occurrence by E'. Hence $\vdash B_a^x \leftrightarrow B'{}_a^x$ by the induction hypothesis. On the other hand, suppose a free occurrence of x in B does lie within the specified occurrence of E in B. Then $B'{}_a^x$ is obtained from B_a^x by replacing an occurrence of E_a^x by $E'{}_a^x$. Since $X \vdash \forall [E \equiv E']$, then $X \vdash \forall [E_a^x \equiv E'{}_a^x]$ by the $\forall$-elimination rule, and therefore by the induction hypothesis $X \vdash B_a^x \leftrightarrow B'{}_a^x$.

THEOREM II.22. *Any formula of the form $(s = t \wedge A_s^x) \to A_t^x$ is a theorem of the ε-calculus.*

Proof. Let n be the number of free occurrences of x in A. Then A_t^x is obtained from A_s^x by replacing n occurrences of s in A_s^x by t. Since $s = t \vdash s = t$, then by n applications of the equivalence rule and the tautology rule we have $s = t \vdash A_s^x \leftrightarrow A_t^x$. (If $n = 0$, then $A_s^x \leftrightarrow A_t^x$ is a tautology.) Consequently, $\vdash s = t \to (A_s^x \leftrightarrow A_t^x)$ by the $\to$-introduction rule, and $\vdash (s = t \wedge A_s^x) \to A_t^x$ by the tautology rule.

7.4 Rule of relabelling bound variables

Let A be any formula. If some well-formed part of A of the form $\varepsilon x B$, $\exists x B$, or $\forall x B$ is replaced by $\varepsilon y [B]_y^x$, $\exists y [B]_y^x$, or $\forall y [B]_y^x$, respectively, where y is not free in B and is free for x in B, then the resulting formula is said to be obtained from A by an *admissible relabelling of a bound variable*. A formula A' is said to be a *variant* of A if there exists a sequence $A_1, \ldots, A_n$ of formulae such that A_1 is A, A_n is A' and for each $i = 2, \ldots, n$, A_i is obtained from A_{i-1} by an admissible relabelling of a bound variable.

THEOREM II.23 *If A' is a variant of A, then $\vdash A \leftrightarrow A'$.*

Proof. Clearly, it is sufficient to prove the theorem for the case where A' is obtained from A by a single admissible relabelling of a bound variable. In this case $\vdash A \leftrightarrow A'$ by the equivalence rule, since by Theorem II.20(i) $\vdash\forall[\varepsilon xB = \varepsilon y[B]_y^x]$, $\vdash\forall[\exists xB \leftrightarrow \exists y[B]_y^x]$, and $\vdash\forall[\forall xB \leftrightarrow \forall y[B]_y^x]$, if y is free for x in B and not free in B.

COROLLARY (The rule of relabelling bound variables). *If $X \vdash A$ and A' is a variant of A, then $X \vdash A'$.*

8 Prenex formulae

A *prenex formula A* is a formula of the form $Q_1 x_1 \ldots Q_n x_n B$ where $n \geqslant 0$, each Q_i is either $\exists$ or $\forall$, the x_i are all distinct variables, and B is an elementary quasi-formula. (Recall that a quasi-formula is elementary if the symbols $\forall$, $\exists$, and ε do not occur in it.) The expression $Q_1 x_1 \ldots Q_n x_n$ is called the *prefix* of A and the quasi-formula B is called the *matrix* of A. Since our definition of a prenex formula includes the possibility that the prefix is empty, it follows that any elementary formula A is a prenex formula, and in this case the matrix of A is A itself.

THEOREM II.24. *For any ε-free formula A, there exists a prenex formula A' such that $\vdash A \leftrightarrow A'$.*

Proof. We first convert A into a formula A_1 which contains no occurrences of the symbols $\wedge$ or $\rightarrow$. This can be done by replacing those quasi-formulae in A of the form $B \wedge C$ by $\neg(\neg B \vee \neg C)$ and those of the form $B \rightarrow C$ by $\neg B \vee C$. Since $\vdash\forall[(B \wedge C) \leftrightarrow \neg(\neg B \vee \neg C)]$ and $\vdash\forall[(B \rightarrow C) \leftrightarrow (\neg B \vee C)]$ by the Tautology Theorem and Theorem II.19(ii), then $\vdash A \leftrightarrow A_1$ by repeated applications of the equivalence rule. We can now convert A_1 into a prenex formula A' by successively replacing those quasi-formulae in A_1 of the form $\neg QxB$ by $Q'x\neg B$ and those of the form $QxB \vee C$ or $C \vee QxB$ by $Qy(B_y^x \vee C)$ or $Qy(C \vee B_y^x)$, respectively, where y is some variable which does not occur free in B or C and which is free for x in B. By Theorem II.20 and by repeated applications of the equivalence rule, we have $\vdash A_1 \leftrightarrow A'$. Hence $\vdash A \leftrightarrow A'$ by the tautology rule.

EXERCISE

Let A be the formula $\forall x\exists yPxy \rightarrow \exists y\forall xPxy$. Convert A into a prenex formula A' such that $\vdash A \leftrightarrow A'$ and the prefix of A' is $\exists x_1\exists x_2\forall x_3\forall x_4$.

9 The addition of new function symbols

We know by the $\forall$-elimination rule and the generalization rule that

$$\vdash \forall yB \quad \text{iff} \quad X \vdash B_a^y,$$

provided that a does not occur in B or in any member of X. The following theorem is an interesting and useful extension of this result.

THEOREM II.25. *Let X be any set of formulae of $\mathscr{L}(\mathscr{V})$ and A any formula of $\mathscr{L}(\mathscr{V})$ of the form $\exists x_1 \ldots \exists x_n \forall y B$, where $n \geq 0$ and the variables $x_1, \ldots, x_n$, and y are all distinct. Let $\mathscr{V}_1$ be the vocabulary obtained from $\mathscr{V}$ by adjoining a new n-place function symbol g. Then*

$$X \vdash_{\varepsilon(\mathscr{V})} \exists x_1 \ldots \exists x_n \forall y B \quad iff \quad X \vdash_{\varepsilon(\mathscr{V}_1)} \exists x_1 \ldots \exists x_n [B]_{gx_1 \ldots x_n}^y.$$

Proof. To simplify the notation we shall write $\vdash$ instead of $\vdash_{\varepsilon(\mathscr{V})}$ and $\vdash_1$ instead of $\vdash_{\varepsilon(\mathscr{V}_1)}$. As a further notational simplification we shall prove the theorem for the case where $n = 1$. However, the method of proof is completely general.

First, assume $X \vdash \exists x \forall y B$. Since every formula of $\mathscr{L}(\mathscr{V})$ is a formula of $\mathscr{L}(\mathscr{V}_1)$ and every axiom of $\varepsilon(\mathscr{V})$ is an axiom of $\varepsilon(\mathscr{V}_1)$, then $X \vdash_1 \exists x \forall y B$. By the $\exists$-elimination rule this yields $X \vdash_1 \forall y [B]_s^x$ where s is $\varepsilon x \forall y B$, and by the $\forall$-elimination rule we then get $X \vdash_1 B_{s\ gs}^{x\ y}$. However, $B_{s\ gs}^{x\ y}$ is $B_{gx\ s}^{y\ x}$. Hence the $\exists$-introduction rule yields $X \vdash \exists x [B]_{gx}^y$.

Secondly, let $\langle A_1, \ldots, A_n \rangle$ be a deduction of $\exists x [B]_{gx}^y$ from X in $\varepsilon(\mathscr{V}_1)$. Let $\varepsilon y' \neg B'$ be a quasi-term obtained from $\varepsilon y \neg B$ by replacing each bound variable in $\varepsilon y \neg B$ by some variable which does not occur (either free or bound) in any of the A_i. For each i, let A_i' be the formula obtained from A_i by replacing each quasi-term in A_i of the form gt, for some t, by the quasi-term $[\varepsilon y' \neg B']_t^x$. (Note that by our relabelling of the bound variables in $\varepsilon y \neg B$, t is free for x in $\varepsilon y' \neg B'$.) We can now prove by induction that for each i, $X \vdash A_i'$. For, if A_i is a member of X, then A_i' is A_i. If A_i is an axiom of $\varepsilon(\mathscr{V}_1)$, other than an E1-axiom, then A_i' is an axiom of $\varepsilon(\mathscr{V})$. If A_i is an E1-axiom, then A_i' is a formula of $\mathscr{L}(\mathscr{V})$ of the same form and therefore a theorem of $\varepsilon(\mathscr{V})$ by virtue of Theorem II.22. Finally, if A_i follows by modus ponens, then so also does A_i'. Since A_n' is the formula $\exists x [B]_{\varepsilon y' \neg B'}^y$, we have $X \vdash \exists x [B]_{\varepsilon y' \neg B'}^y$. The rule of relabelling bound variables now yields $X \vdash \exists x [B]_{\varepsilon y \neg B}^y$, and the desired result $X \vdash \exists x \forall y B$ follows by applications of the $\exists$-elimination rule, $\forall$-introduction, and $\exists$-introduction rules. (*Note:* it is implicit in the statement of the theorem that gx is free for y in B. Hence $\varepsilon y \neg B$ is free for y in B, and therefore for any term s, $[B]_{\varepsilon y \neg B\ s}^{y\ x}$ is $[B]_{s\ \varepsilon y \neg [B]_s^x}^{x\ y}$.)

THEOREM II.26. *Let X be any set of formulae of $\mathscr{L}(\mathscr{V})$, C any formula of $\mathscr{L}(\mathscr{V})$, and A any formula of $\mathscr{L}(\mathscr{V})$ of the form $\forall x_1 \ldots \forall x_n \exists y B$, where $n \geq 0$ and the variables $x_1, \ldots, x_n$ and y are all distinct. Let $\mathscr{V}_1$ be the vocabulary obtained from $\mathscr{V}$ by adjoining a new n-place function symbol g. Then*

$$X, \forall x_1 \ldots \forall x_n \exists y B \vdash_{\varepsilon(\mathscr{V})} C \quad iff \quad X, \forall x_1 \ldots \forall x_n [B]_{gx_1 \ldots x_n}^y \vdash_{\varepsilon(\mathscr{V}_1)} C.$$

Proof. Using the same notational simplifications as above, we want to prove

$$X, \forall x \exists y B \vdash C \quad \text{iff} \quad X, \forall x [B]_{gx}^{y} \vdash_1 C.$$

By Theorem II.25, we have

$$X, \neg C \vdash \exists x \forall y \neg B \quad \text{iff} \quad X, \neg C \vdash_1 \exists x \neg [B]_{gx}^{y}.$$

However, Theorem II.20(ii) and the equivalence rule yield

$$\vdash \exists x \forall y \neg B \leftrightarrow \neg \forall x \exists y B$$

and

$$\vdash \exists x \neg [B]_{gx}^{y} \leftrightarrow \neg \forall x [B]_{gx}^{y}.$$

Hence, by the tautology rule

$$X, \neg C \vdash \neg \forall x \exists y B \quad \text{iff} \quad X, \neg C \vdash_1 \neg \forall x [B]_{gx}^{y}.$$

The desired result now follows by the $\rightarrow$-elimination, $\rightarrow$-introduction, and contrapositive rules.

Theorem II.26 provides a formal justification in terms of the ε-calculus of a type of reasoning which is commonly used in mathematics. Suppose that we are trying to prove some statement C and in the course of the proof we prove a statement of the form

(1)   'for all x, there exists a y such that $B(x,y)$'

where $B(x,y)$ asserts some relationship between x and y. It is often convenient to have, for each x, a way of denoting some y such that $B(x,y)$. Since the notation must express the fact that y depends on x, we introduce a new function symbol g and say

(2)   'for all x, $B(x,g(x))$'

thus using $g(x)$ to denote an appropriate y. Theorem II.26 shows that if we can deduce C using statement (2), then we can deduce C directly from statement (1) without using the function g. Of course our justification of this line of reasoning makes use of the ε-symbol and the logical power of the ε-calculus, since in effect what we have done is to identify the expression $g(x)$ with the quasi ε-term $\varepsilon y B(x,y)$ and to use the fact that under this identification statements (1) and (2) are equivalent in the logic of the ε-calculus. However, once we have proved the Second ε-Theorem, it will follow that the above line of reasoning is justifiable even when one is using a more standard system of logic (i.e., the predicate calculus) which does not include the ε-symbol. (See Chapter III, §4.1.)

9.1 Skolem and Herbrand resolutions of prenex formulae

A prenex formula is ∃-*prenex* if the symbol ∀ does not occur in its prefix and ∀-*prenex* if the symbol ∃ does not occur in its prefix. In this section we shall assign to any prenex formula A a certain ∃-prenex formula, denoted by A_H, and a certain ∀-prenex formula, denoted by A_S. The formula A_H is called the *Herbrand resolution* of A, and the formula A_S the *Skolem resolution* of A. Using Theorems II.25 and II.26, we can establish close relationships between A_H and A and between A_S and A.

Let $\mathcal{V}$ be any vocabulary and A any prenex formula of $\mathcal{L}(\mathcal{V})$. The *Herbrand resolution*, A_H, of A is defined as follows. If A is already $\exists$-prenex, then A_H is A. Otherwise, A is of the form $\exists x_1 \ldots \exists x_n \forall y B$, where $n \geq 0$. Let $\mathcal{V}'$ be the vocabulary obtained by adjoining a new n-place function symbol g to $\mathcal{V}$. Let A' be the formula $\exists x_1 \ldots \exists x_n [B]^y_{g x_1 \ldots x_n}$. If A' is $\exists$-prenex, then A_H is A'. Otherwise, repeat the procedure by adding a new function symbol to $\mathcal{V}'$ and forming the prenex formula A''. After a finite number of steps a $\exists$-prenex formula is obtained. This formula is A_H. The new function symbols which are added to $\mathcal{V}$ in forming A_H are called *Herbrand functions*.

The Skolem resolution, A_S, of A is defined similarly. (Thus if A is of the form $\forall x_1 \ldots \forall x_n \exists y B$, then A' is the formula $\forall x_1 \ldots \forall x_n [B]^y_{g x_1 \ldots x_n}$.) The new function symbols which are added to $\mathcal{V}$ is forming A_S are called *Skolem functions*. The process whereby A is converted to A_S is often referred to as *symbolic resolution* of existential formulae.

For example, suppose A is the prenex formula

$$\forall x_1 \exists y_1 \forall x_2 \exists y_2 B,$$

where B is the matrix of A. Then the formula

$$\exists y_1 \exists y_2 [B]^{x_1 \ x_2}_{g_1 \ g_2 y_1}$$

is the Herbrand resolution of A, where g_1 and g_2 are used as Herbrand functions, and the formula

$$\forall x_1 \forall x_2 [B]^{y_1 \quad y_2}_{h_1 x_1 \ h_2 x_1 x_2}$$

is the Skolem resolution of A, where h_1 and h_2 are used as Skolem functions.

THEOREM II.27. *Let A be any prenex formula of $\mathcal{L}(\mathcal{V})$, X any set of formulae of $\mathcal{L}(\mathcal{V})$, and $\mathcal{V}^*$ the vocabulary obtained from $\mathcal{V}$ by adjoining the Herbrand functions used in forming A_H. Then $X \vdash_{\varepsilon(\mathcal{V})} A$ iff $X \vdash_{\varepsilon(\mathcal{V}^*)} A_H$.*

Proof. The proof follows from Theorem II.25 by induction on the number of occurrences of $\forall$ in the prefix of A.

THEOREM II.28. *Let A be any prenex formula of $\mathcal{L}(\mathcal{V})$, C any formula of $\mathcal{L}(\mathcal{V})$, X any set of formulae of $\mathcal{L}(\mathcal{V})$, and $\mathcal{V}^*$ the vocabulary obtained from $\mathcal{V}$ by adjoining the Skolem functions used in forming A_S. Then*

$$X, A \vdash_{\varepsilon(\mathcal{V})} C \quad iff \quad X, A_S \vdash_{\varepsilon(\mathcal{V}^*)} C.$$

Proof. The proof follows from Theorem II.26 by induction on the number of occurrences of $\exists$ in the prefix of A.

EXERCISE

Let A be any prenex formula of $\mathscr{L}(\mathscr{V})$. Prove (i) $\vdash_{\varepsilon(\mathscr{V}*)} A \to A_H$, where $\mathscr{V}*$ is as in Theorem II.27, and (ii) $\vdash_{\varepsilon(\mathscr{V}*)} A_S \to A$, where $\mathscr{V}*$ is as in Theorem II.28. Find a prenex formula A such that $A_H \to A$ is invalid and therefore not a theorem of $\varepsilon(\mathscr{V}*)$.

10   The elementary calculus

Let $\mathscr{V}$ be any vocabulary. We shall now define two formal systems for $\mathscr{V}$, the *elementary calculus* and the *elementary calculus without identity*. The first of these will be denoted by $EC(\mathscr{V})$ and the second by $EC'(\mathscr{V})$. Each may be regarded as a subsystem of $\varepsilon(\mathscr{V})$.

Recall that a formula of $\mathscr{L}(\mathscr{V})$ is *elementary* if it does not contain the symbols $\forall$, $\exists$, or ε. (Equivalently, a formula is elementary if it contains no variables.) The *axioms* of $EC(\mathscr{V})$ are the elementary formulae of $\mathscr{L}(\mathscr{V})$ which are instances of axiom schemata P1–P10, E1, and E3. For any set X of formulae of $\mathscr{L}(\mathscr{V})$ and any formula A of $\mathscr{L}(\mathscr{V})$, a *deduction* of A from X in $EC(\mathscr{V})$ is any sequence $\langle A_1, \ldots, A_n \rangle$ of elementary formulae of $\mathscr{L}(\mathscr{V})$ such that A_n is A and for each $i = 1, \ldots, n$, A_i is an axiom of $EC(\mathscr{V})$, or A_i is a member of X, or A_i follows by modus ponens from A_j and A_k for some $j, k < i$. In other words a sequence of formulae of $\mathscr{L}(\mathscr{V})$ is a deduction in $EC(\mathscr{V})$ if and only if it is a deduction in $\varepsilon(\mathscr{V})$ and each of its members is elementary. As usual, we write $X \vdash_{EC(\mathscr{V})} A$ to denote that there exists a deduction of A from X in $EC(\mathscr{V})$. Notice that if $X \vdash_{EC(\mathscr{V})} A$, then A must be elementary. Since any deduction in $EC(\mathscr{V})$ is automatically a deduction in $\varepsilon(\mathscr{V})$, then $X \vdash_{EC(\mathscr{V})} A$ implies $X \vdash_{\varepsilon(\mathscr{V})} A$. We shall see later that the converse is also true, i.e., $X \vdash_{\varepsilon(\mathscr{V})} A$ implies $X \vdash_{EC(\mathscr{V})} A$, provided that X is a set of elementary formulae and A is elementary (see page 64).

The formal system $EC'(\mathscr{V})$ is defined in a similar way except that the axioms of $EC'(\mathscr{V})$ do not include instances of E1 and E3, and a deduction in $EC'(\mathscr{V})$ must be a sequence of elementary, *identity-free* formulae.

Since, as usual, $\mathscr{V}$ is an arbitrary vocabulary, we shall write EC instead of $EC(\mathscr{V})$ and EC' instead of $EC'(\mathscr{V})$.

Many of the derived rules of inference which we have proved for the ε-calculus also apply to the elementary calculus (without identity). In particular we can prove exactly as for the ε-calculus that the $/$-rule, contradiction rule, $\neg\neg$-rule, α-rule, and β-rule all hold in EC and EC' (assuming, of course, that the letters A, α, and β which appear in the statements of these rules now denote elementary (identity-free) formulae). Consequently, the proofs of Theorems II.13 and II.14 (the Tautology Theorem) can be used verbatim to prove the following theorem.

THEOREM II.29. *For any elementary (identity-free) formula A, if A is a tautology, then A is a theorem of the elementary calculus (without identity).*

A theorem of the elementary calculus is not necessarily a tautology, since for example, $a = a$ is a theorem but not a tautology. Nonetheless, we have the following very useful theorem.

THEOREM II.30. *Every identity-free theorem of EC is a tautology.*

Proof. For any formula B, we write $h(B)$ to denote the formula which is obtained from B by replacing every occurrence in B of a quasi-formula of the form $s = t$ by f if s and t are not the same quasi-terms and by $\neg f$ if they are the same.

Now let $\langle A_1, \ldots, A_n \rangle$ be a proof in EC of some identity-free formula A. We shall prove by induction that for each A_i, $h(A_i)$ is a tautology, thus proving that $h(A_n)$, i.e. A, is a tautology.

Case 1. A_i is a propositional axiom: Then $h(A_i)$ is also a propositional axiom and therefore a tautology by Exercise 2, page 41.

Case 2. A_i is an E3-axiom: Then $h(A_i)$ is $\neg f$ which is a tautology.

Case 3. A_i is an E1-axiom: Thus A_i has the form $(s = t \wedge B_s^x) \to B_t^x$. If s and t are not the same terms, then $h(A_i)$ has the form $(f \wedge C_1) \to C_2$, which is a tautology. On the other hand, if s and t are the same terms, then $h(A_i)$ has the form $(\neg f \wedge C) \to C$, which is also a tautology.

Case 4. A_i follows by modus ponens from A_j and A_k, where $j, k < i$: Then $h(A_i)$ follows from $h(A_j)$ and $h(A_k)$ by modus ponens. Hence by the induction hypothesis and Exercise 2, page 18, $h(A_i)$ is a tautology.

Theorems II.29 and II.30 together imply that for any identity-free formula A, A is a tautology iff $\vdash_{EC} A$. Furthermore these theorems provide the interesting, though perhaps not unexpected, information that any identity-free theorem of EC is a theorem of EC'. In other words, in the elementary calculus the identity symbol and the equality axioms are superfluous in proving identity-free theorems. We shall return to the general problem of proving the eliminability of the identity symbol in Chapter III, page 83.

11 The predicate calculus

We now turn to the standard formalization of logic, the predicate calculus. For any vocabulary $\mathscr{V}$, the predicate calculus for $\mathscr{V}$, which we shall denote by PC($\mathscr{V}$) will have as its deductions certain sequences of ε-free formulae. We want to define the notion of a deduction in PC($\mathscr{V}$) in such a way that all the derived rules of inference which hold for the ε-calculus hold also for the predicate calculus, except of course those rules, such as the ∃-elimination and ∀-introduction rules, in which certain ε-terms are specified. Since the quantifier rules, such as the rule of generalization, ∃-rule, and ∀-rule, depend on the axiom schema

Q4 $\exists x A \rightarrow A(\varepsilon x A)$

and since instances of this schema cannot be used in deductions in the predicate calculus, we compensate for this deficiency by adopting the ∃-rule as one of the basic rules of inference of the system. Formally, we state this rule as follows:

> A formula of the form $\exists x A \rightarrow B$ *follows by the* ∃-*rule* from a formula C, if and only if C is of the form $A_a^x \rightarrow B$, where a does not occur in A or B.

The *axioms* of the predicate calculus for $\mathscr{V}$ are the ε-free instances of the axiom schemata P1–P10, Q1–Q3, E1, and E3. Thus the axioms of PC($\mathscr{V}$) are simply the ε-free axioms of $\varepsilon(\mathscr{V})$.

In defining a deduction in PC($\mathscr{V}$) we are faced with the following problem. Suppose we were to proceed in the usual fashion by defining a deduction of A from X as any sequence $\langle A_1, \ldots, A_n \rangle$ of ε-free formulae of $\mathscr{L}(\mathscr{V})$, such that A_n is A and for each i, A_i is an axiom, or A_i is a member of X, or A_i follows by modus ponens or by the ∃-rule. Unfortunately, under this definition the →-introduction rule (Deduction Theorem) does not hold in its full generality (cf. Mendelson [1964], pp. 60–61). One standard way of overcoming this deficiency is to stipulate that in each application of the ∃-rule the specified symbol a must not occur in any member of X. However, this restriction is too strong, for although the →-introduction rule now holds, we lose the MP rule as well as the simple rule which asserts that if $X \subseteq Y$ and $X \vdash A$, then $Y \vdash A$. Consequently, we adopt the following modified restriction on the ∃-rule. (Our definition of a deduction is due to Lyndon [1966]. For other suitable definitions as well as a full analysis of the problem see Montague and Henkin [1956].)

First of all we define a 'derivation' in PC($\mathscr{V}$). For any finite set Y of formulae of $\mathscr{L}(\mathscr{V})$ and any formula A of $\mathscr{L}(\mathscr{V})$ a *derivation* of A from Y in PC($\mathscr{V}$) is any sequence $\langle A_1, \ldots, A_n \rangle$ of ε-free formulae of $\mathscr{L}(\mathscr{V})$, where A_n is A and for each $i = 1, \ldots, n$ at least one of the following conditions holds:

(i) A_i is an axiom of PC($\mathscr{V}$),

(ii) A_i is a member of Y,

(iii) A_i follows by modus ponens from A_j and A_k for some $j, k < i$,

(iv) A_i follows by the ∃-rule from some A_j, where $j < i$, provided that the specified individual symbol in A_j does not occur in any member of Y.

(Notice that this definition of a derivation corresponds to the second unsuccessful definition of a deduction given above.) We now define a *deduction* of A from X in PC($\mathscr{V}$) as any derivation of A from some finite subset Y of X. As usual, if there exists a deduction of A from X in PC($\mathscr{V}$) we write $X \vdash_{\text{PC}(\mathscr{V})} A$, or simply $X \vdash_{\text{PC}} A$ or $X \vdash A$ when there is no possibility of ambiguity. Notice that by defining a deduction in this way we still have the rule which asserts that if $X \subseteq Y$ and $X \vdash A$, then $Y \vdash A$.

11.1   Derived rules of inference for the predicate calculus

Some of the derived rules of inference, such as the MP rule, $\rightarrow$-elimination rule, and $\rightarrow$-introduction rule, which were so easy to verify for the ε-calculus, are more difficult to establish in the case of the predicate calculus. For example, the proof of the $\rightarrow$-elimination rule is complicated by the fact that if $\langle A_1, \ldots, A_n \rangle$ is a derivation of $A \rightarrow B$ from X in PC, then the sequence $\langle A_1, \ldots, A_n, A, B \rangle$ is not necessarily a derivation of B from $X \cup \{A\}$ since an individual symbol occurring in A may be involved in an application of the $\exists$-rule. To overcome this difficulty we need the following theorem.

THEOREM II.31. *If X and Y are any finite sets of formulae and $\mathscr{D}$ is a derivation of A from X, then there exists a derivation $\mathscr{D}'$ of A from $X \cup Y$.*

Proof. Let I be the (finite) collection of individual symbols which occur in the members of Y. We want to prove that there exists a derivation $\mathscr{D}'$ of A from X in which no member of I is used in an application of the $\exists$-rule, since then $\mathscr{D}'$ is a derivation of A from $X \cup Y$. The proof is by induction on the number m of members of I which are so used in $\mathscr{D}$. If $m = 0$, then $\mathscr{D}$ itself is the required derivation. Suppose $m > 0$. Let a be some member of I which is involved in an application of the $\exists$-rule in $\mathscr{D}$ and let b be an individual symbol not in I and not occurring in any member of X or any member of $\mathscr{D}$. Suppose $\mathscr{D}$ is the sequence $\langle A_1, \ldots, A_n \rangle$. Let $\mathscr{D}'$ be the sequence $\langle A_1', \ldots, A'_{n-1}, A_1, \ldots, A_n \rangle$, where each A_i' is obtained from A_i by replacing every occurrence of a by b. Then $\mathscr{D}'$ is a derivation of A from X in which only $m - 1$ members of I are involved in applications of the $\exists$-rule. For, if A_i follows from A_j in $\mathscr{D}$ by an application of the $\exists$-rule which involves a, then the presence of A_i in $\mathscr{D}'$ can be justified by applying the $\exists$-rule to A_j'. This application of the $\exists$-rule involves the new symbol b. The desired result now follows by induction.

Using this result we can prove the counterpart of Theorem II.3(ii). (The counterparts of parts (i), (iii), and (iv) are trivial.)

THEOREM II.32. *Let X and Y be any sets of formulae and $A, B_1, \ldots, B_n$ any ε-free formulae. If $Y, B_1, \ldots, B_n \vdash A$ and $X \vdash B_i$ for each $i = 1, \ldots, n$, then $X \cup Y \vdash A$.*

Proof. It is sufficient to consider the case where $n = 1$. Thus we want to prove that if $Y, B \vdash A$ and $X \vdash B$, then $X \cup Y \vdash A$. Let $\mathscr{D}_1$ be a derivation of A from Y' and $\mathscr{D}_2$ a derivation of B from X', where Y' is a finite subset of $Y \cup \{B\}$ and X' is a finite subset of X. By Theorem II.31, there exist derivations $\langle A_1, \ldots, A_m \rangle$ of B from $X' \cup Y'$ and $\langle B_1, \ldots, B_n \rangle$ of A from $X' \cup Y'$.

Hence the sequence $\langle A_1, \ldots, A_m, B_1, \ldots, B_n \rangle$ is a derivation of A from $X' \cup (Y' \setminus \{B\})$ and therefore a deduction of A from $X \cup Y$.

EXERCISE

Prove that the $\rightarrow$-elimination rule and MP rule hold for the predicate calculus, where of course the letters A, B, B_1, ..., B_n now denote ε-free formulae.

THEOREM II.33. *For any ε-free formula A, if A is a tautology, then $\vdash_{PC} A$.*

Proof. For the purposes of this proof we write $PC_0(\mathscr{V})$ to denote the formal system which is obtained from $PC(\mathscr{V})$ by excluding the $\exists$-rule. Thus the deductions in $PC_0(\mathscr{V})$ are simply the deductions in $\varepsilon(\mathscr{V})$ in which every formula is ε-free. For this formal system one can prove the Tautology Theorem just as it was proved for the ε-calculus. Hence if A is an ε-free tautology, then $\vdash_{PC_0} A$ and *a fortiori* $\vdash_{PC} A$.

The tautology rule now follows immediately using this theorem and the MP rule.

We shall use the following important result in the next chapter to prove the Second ε-Theorem.

THEOREM II.34. *If $\vdash (\exists x A \rightarrow A_a^x) \rightarrow C$, where a does not occur in A or C, then $\vdash C$.*

Proof. Assume $\vdash (\exists x A \rightarrow A_a^x) \rightarrow C$. Since $A_a^x \rightarrow C$ is a tautological consequence of $(\exists x A \rightarrow A_a^x) \rightarrow C$, we have $\vdash A_a^x \rightarrow C$ by the tautology rule, and hence $\vdash \exists x A \rightarrow C$ by an application of the $\exists$-rule. However, C is a tautological consequence of $(\exists x A \rightarrow A_a^x) \rightarrow C$ and $\exists x A \rightarrow C$. Hence $\vdash C$ by the tautology rule.

EXERCISES

1. Prove that the $\rightarrow$-introduction rule (Deduction Theorem) holds for the predicate calculus.

2. Show that the counterpart of Theorem II.17 holds for the predicate calculus.

11.2   The predicate calculus without identity

If there exists a deduction of A from X in $PC(\mathscr{V})$ in which every formula is identity-free we write $X \vdash_{PC'(\mathscr{V})} A$. Thus $PC'(\mathscr{V})$ may be regarded as a formal system for $\mathscr{V}$ whose axioms are the identity-free axioms of $PC(\mathscr{V})$ and whose deductions are the identity-free deductions of $PC(\mathscr{V})$. This formal system for $\mathscr{V}$ is called the *predicate calculus without identity*. Obviously, all the results which we proved for PC($\mathscr{V}$) in the last section also hold for PC'($\mathscr{V}$), assuming that we restrict our attention to identity-free formulae.

In the next chapter we shall prove that for any identity-free, ε-free formula A, if $\vdash_{PC} A$, then $\vdash_{PC'} A$. Notice that the technique which we used to prove the analogous result for the elementary calculus cannot be used for the case of the predicate calculus. For, suppose A is the axiom $\neg \exists x(x = a) \rightarrow \neg a = a$. Then using the technique employed in proving Theorem II.30, the identity-free formula which is assigned to A is the formula $\neg \exists x f \rightarrow \neg \neg f$. This latter formula is not a theorem of PC'.

EXERCISE

Prove that the formula $\neg \exists x f \rightarrow \neg \neg f$ is not a theorem of PC'. (*Hint*: use the technique employed in proving Theorem II.15.)

12   The formal superiority of the ε-calculus

It should now be apparent that the ε-calculus is a much simpler and neater formalization of logic than is the predicate calculus.

The deductions in the ε-calculus can be defined in a simple straightforward way, and the basic derived rules of inference can be established with very little difficulty. On the other hand, we have seen that no matter how one defines a deduction in the predicate calculus certain complications arise since one must adopt an additional rule of inference for dealing with the quantifiers.

Furthermore, in the ε-calculus the derived rules of inference for the quantifiers can be expressed more easily and can be used more conveniently than in the predicate calculus. For example, although it is possible to formulate a derived rule of inference for the predicate calculus which is analogous to the $\exists$-introduction rule (cf. rule C in Mendelson [1964], page 74), this rule is subject to many tedious restrictions.

In spite of these formal advantages which are gained by using the ε-symbol, the question arises whether the use of such an indeterminate logical symbol is philosophically justified. In the next chapter we shall give the best possible justification for the use of the ε-symbol in logic by proving that any ε-free theorem of the ε-calculus is a theorem of the predicate calculus. In other words, the ε-symbol and the axioms associated with it can be eliminated from proofs of ε-free formulae.

CHAPTER III

THE ε-THEOREMS

1 Introduction

In Chapter II we defined three basic formal systems (for a given vocabulary $\mathscr{V}$): the elementary calculus, the predicate calculus, and the ε-calculus. Loosely speaking, the predicate calculus is obtained from the elementary calculus by adjoining the quantifiers and the appropriate axioms and rule of inference for dealing with them, and the ε-calculus is obtained from the predicate calculus by adjoining the ε-symbol and the appropriate axioms for dealing with it (and by excluding the redundant ∃-rule). Thus by passing from the elementary calculus to the predicate calculus and from the predicate calculus to the ε-calculus we obtain stronger and stronger logical systems. One of the main objectives of this chapter is to show that these successive strengthenings of the elementary calculus are completely justified since in each case the new logical symbols and axioms are in a certain sense eliminable. To put it more precisely, we shall prove the following two theorems.

1. If A is an elementary formula of $\mathscr{L}(\mathscr{V})$, X is a set of elementary formulae of $\mathscr{L}(\mathscr{V})$, and $X \vdash_{PC(\mathscr{V})} A$, then $X \vdash_{EC(\mathscr{V})} A$.
2. If A is an ε-free formula of $\mathscr{L}(\mathscr{V})$, X is a set of ε-free formulae of $\mathscr{L}(\mathscr{V})$, and $X \vdash_{\varepsilon(\mathscr{V})} A$, then $X \vdash_{PC(\mathscr{V})} A$.

The second of these two statements is known as Hilbert's Second ε-Theorem and the first is a special case of his First ε-Theorem. The full statement of the First ε-Theorem is as follows.

1′. If A is any prenex formula of $\mathscr{L}(\mathscr{V})$, X is any set of prenex formulae of $\mathscr{L}(\mathscr{V})$, and $X \vdash_{PC(\mathscr{V})} A$, then $Z \vdash_{EC(\mathscr{V})} B_1 \vee \ldots \vee B_n$, where each member of Z is some substitution instance of the matrix of some member of X and each of the B_i is a substitution instance of the matrix of A.

We shall also see in this chapter that besides providing a formal justification for these extensions of the elementary calculus, the two ε-theorems can be used to prove some important results about the predicate calculus, such as Skolem's Theorem (Theorem III.12) and Herbrand's Theorem (Theorem III.14).

As in Chapter II, our proofs will be completely finitary in the sense that whenever we prove that a certain deduction exists, our method of proof will provide a technique for constructing such a deduction. It should be pointed out that by using a non-constructive, model-theoretic argument the Second ε-Theorem follows trivially from the completeness of the predicate calculus. For if $X \vdash_\varepsilon A$, where X and A are ε-free, then $X \vDash A$ by the soundness of the ε-calculus (cf. Exercise 3, page 41), and hence $X \vdash_{PC} A$ by the completeness of the predicate calculus.

Throughout this chapter the vocabulary $\mathcal{V}$ shall be some arbitrary, but fixed vocabulary.

2.1 The basic problem

Since ε-terms play the same role in the ε-calculus as individual symbols play in the predicate calculus, it would seem that one could prove the Second ε-Theorem by assigning an appropriate individual symbol to each ε-term in a deduction $\mathcal{D}$ and then forming a new deduction $\mathcal{D}'$ by simply replacing each ε-term by its assigned individual symbol and making the necessary alterations so that the $\exists$-rule can be used where the Q4-axioms were used in $\mathcal{D}$.

However, this procedure is complicated by the following problem. Suppose that A_i is an axiom of the ε-calculus, εyB is some ε-term, and $A_i{}'$ is the formula obtained from A_i by replacing each occurrence of εyB by some term s. Then $A_i{}'$ may fail to be an axiom, namely if one of the following cases holds:

(i) A_i is the Q4-axiom $\exists yB \rightarrow B(\varepsilon yB)$, in which case $A_i{}'$ is $\exists yB \rightarrow B(s)$;

(ii) A_i is an E2-axiom in which εyB is one of the two specified ε-terms;

(iii) A_i is a Q3-axiom or Q4-axiom and an occurrence of the specified term in that axiom (i.e., t or εxA) lies within an occurrence of εyB.

The third case is the most troublesome one. Consider the following example. Suppose A_i is the Q3-axiom

(1) $\neg \exists x(x = \varepsilon y(y = x)) \rightarrow \neg(t = \varepsilon y(y = t))$

and εyB is the term $\varepsilon y(y = t)$. Notice that an occurrence of the specified term t of this Q3-axiom lies within an occurrence of $\varepsilon y(y = t)$. Now if we replace every occurrence of $\varepsilon y(y = t)$ in A_i by the term s, we obtain the formula

(2) $\neg \exists x(x = \varepsilon y(y = x)) \rightarrow \neg(t = s)$.

This formula is not an instance of axiom schema Q3.

Notice that case (iii) can only arise if A_i is of the form $\neg \exists xA \rightarrow \neg A(t)$ or $\exists xA \rightarrow A(\varepsilon xA)$ where a free occurrence of x in A lies within the scope of an ε-symbol. Thus, in the above example, where A is the quasi-formula

$$x = \varepsilon y(y = x),$$

the second free occurrence of x lies within the quasi ε-term $\varepsilon y(y = x)$. Formulae such as (1) above are examples of what we shall now call 'improper' formulae.

2.2   Proper and improper formulae

Let A be any term or formula. If t is a term which occurs in A, but is not A itself, then t is called a *subterm* of A. For any formula A, the *skeleton* of A is that formula which is obtained from A by successively replacing each subterm of A by the individual symbol a_1 (starting with those subterms of maximal length). We shall denote the skeleton of A by A^+. A formula A is *proper* if its skeleton is ε-free, and *improper* if its skeleton is not ε-free.

For example, suppose A is the formula

$$\neg \exists x(x = \varepsilon y(y = s)) \rightarrow \neg(t = \varepsilon y(y = s)),$$

where s and t are any two terms. Then the skeleton of A is the formula

$$\neg \exists x(x = a_1) \rightarrow \neg(a_1 = a_1).$$

Since this latter formula is ε-free, then A is proper. On the other hand, suppose A is the formula

$$\neg \exists x(x = \varepsilon y(y = x)) \rightarrow \neg(t = \varepsilon y(y = t)).$$

Then the skeleton of A is the formula

$$\neg \exists x(x = \varepsilon y(y = x)) \rightarrow \neg(a_1 = a_1),$$

and therefore, in this case, A is improper.

Obviously, if A_i is a proper formula of the form $\neg \exists x A \rightarrow \neg A(t)$ or of the form $\exists x A \rightarrow A(\varepsilon x A)$, then x does not have a free occurrence in A within the scope of an ε-symbol. Consequently, if A_i is a proper axiom of the ε-calculus and $A_i{}'$ is obtained from A_i by replacing every occurrence of $\varepsilon y B$ by some term s, then the troublesome case (iii) considered above cannot arise. Therefore, we have the following theorem.

THEOREM III.1. *Let $\varepsilon y B$ be any ε-term, s any term, and A any axiom of the ε-calculus other than an E2-axiom or the Q-axiom $\exists y B \rightarrow B(\varepsilon y B)$. If A is a proper formula, then the formula obtained from A by replacing each occurrence of $\varepsilon y B$ by s is an axiom.*

2.3   The ε*-calculus

We shall now use the new notion of a proper formula to define a certain formal system called the ε*-calculus (for $\mathscr{V}$). We then prove quite easily that the Second ε-Theorem holds for this weakened version of the ε-calculus.

The *axioms* of the ε*-calculus (for $\mathscr{V}$) are all *proper* formulae of $\mathscr{L}(\mathscr{V})$ which are instances of axiom schemata P1–P10, Q1–Q4, E1, and E3. (Note that we exclude E2-axioms.) A *deduction* of A from X in the ε*-calculus (for $\mathscr{V}$) is any sequence $\langle A_1, \ldots, A_n \rangle$ of proper formulae (of $\mathscr{L}(\mathscr{V})$) such that A_n is A and for each $i = 1, \ldots, n$ either A_i is an axiom, or A is a member of

X, or A_i follows by modus ponens from some A_j and A_k, where $j, k < i$. In other words, a deduction in the ε*-calculus is a deduction in the ε-calculus in which every formula is proper and no instances of axiom schema E2 are used as axioms. As usual we write $X \vdash_{\varepsilon*} A$ to denote that there exists a deduction of A from X in the ε*-calculus.

Obviously the arguments used in proving the →-introduction rule, →-elimination rule, and tautology rule for the ε-calculus apply equally well in the case of the ε*-calculus. Furthermore Theorem II.17 can be proved for the ε*-calculus just as it was proved for the ε-calculus, since if A_i is a proper formula and A_i' is the formula obtained from A_i by replacing each occurrence of a by some term t, then A_i' is also proper. Consequently, the ∃-rule holds for the ε*-calculus. In fact, the only derived rules of inference which hold for the ε-calculus but not for the ε*-calculus are those rules, such as the distribution rule, rule of relabelling bound variables, and equivalence rule, which depend on axiom schema E2. Even so, these rules still hold for the ε*-calculus if one restricts one's attention to ε-free formulae.

In view of the fact that the ∃-rule holds for the ε*-calculus, we have the following theorem.

THEOREM III.2. *If $X \vdash_{PC} A$, then $X \vdash_{\varepsilon*} A$.*

Proof. Assume $X \vdash_{PC} A$. Then there exists a derivation $\langle A_1, \ldots, A_n \rangle$ of A from Y in the predicate calculus, where Y is some finite subset of X. It follows trivially by induction that, for each $i = 1, \ldots, n$, $Y \vdash_{\varepsilon*} A_i$. Hence $X \vdash_{\varepsilon*} A$.

We now prove the converse of Theorem III.2. This result may be regarded as a weaker form of the Second ε-Theorem.

THEOREM III.3. *For any ε-free X and A, if $X \vdash_{\varepsilon*} A$, then $X \vdash_{PC} A$.*

By appealing to the →-introduction rule for the ε*-calculus and the →-elimination rule for the predicate calculus, the proof of Theorem III.3 can be reduced to proving the following lemma.

LEMMA. *If $\mathscr{D}$ is a proof in the ε*-calculus of some ε-free formula C, then $\mathscr{D}$ can be converted into a proof of C in the predicate calculus.*

Proof. The proof is by induction on the number n of distinct ε-terms occurring in $\mathscr{D}$. If $n = 0$, then $\mathscr{D}$ is already a proof in the predicate calculus.

Suppose $n > 0$. Let $\varepsilon y B$ be an ε-term occurring in $\mathscr{D}$ such that the length of $\varepsilon y B$ is less than or equal to the length of every other ε-term occurring in $\mathscr{D}$. Hence no ε-terms occur in B. Let a be some individual symbol which does not occur in B or C, and let $\mathscr{D}'$ be the sequence of formulae obtained from $\mathscr{D}$ by replacing each occurrence of $\varepsilon y B$ by a. If the formula

$$(1) \qquad \exists y B \rightarrow B(\varepsilon y B)$$

is not used as an axiom in $\mathscr{D}$, then by Theorem III.1 $\mathscr{D}'$ is a proof of C in the ε*-calculus. Since only n - 1 distinct ε-terms occur in $\mathscr{D}'$, then by the induction hypothesis there exists a proof of C in the predicate calculus. Suppose on the other hand that the formula (1) *is* used as an axiom in $\mathscr{D}$. Then by Theorem III.1, $\mathscr{D}'$ is a deduction in the ε*-calculus of C from the formula

(2) $\exists y B \to B(a).$

By the $\to$-introduction rule $\mathscr{D}'$ can be converted into a proof $\mathscr{D}''$ in the ε*-calculus of the formula

(3) $(\exists y B \to B(a)) \to C.$

Since C is ε-free, B contains no ε-terms, and (1) is proper, it follows that (3) is ε-free. Furthermore only n - 1 distinct ε-terms occur in $\mathscr{D}''$. Consequently, by the induction hypothesis there exists a proof of (3) in the predicate calculus. By our choice of a and by Theorem II.34, this implies that there exists a proof of C in the predicate calculus.

2.4 The usefulness of the ε*-calculus

The ε*-calculus is considerably weaker than the ε-calculus since (i) the E2-axioms are excluded and (ii) only proper formulae may be used in a deduction. Nonetheless, this weaker system can play a very useful role in the study of mathematical logic. For example, in order to prove various results about the predicate calculus, such as the $\to$-introduction rule (Deduction Theorem), generalization rule, substitution rule, distribution rule, equivalence rule, rule of relabelling bound variables, etc., it is perhaps easier to show that these results hold in the ε*-calculus (subject to certain conditions) and then use Theorems III.2 and III.3 to prove that they hold in the predicate calculus.

It has been suggested (Fraenkel and Bar-Hillel [1958] page 184) that the ε-symbol be used in teaching elementary logic. However, if a full analysis of the ε-calculus together with a proof of the Second ε-Theorem seems too ambitious for an introductory course in logic, one could easily restrict one's attention to the ε*-calculus. In this case one would modify the rules of formation of the formal languages by stating that an expression of the form $\varepsilon x A$ is well-formed provided that no variable other than x is free in A. In this way all the formulae of the language would be proper formulae. This is essentially the method which is used by Shoenfield [1967] (page 46) although the ε-symbol is not explicitly mentioned. Shoenfield shows that this approach provides a simple proof of the completeness of the predicate calculus. Furthermore, we shall see in a later section that in order to prove the First ε-Theorem one may use the ε*-calculus and avoid the ε-calculus altogether.

Because of the exclusion of the E2-axioms from the ε*-calculus, the equivalence rule does not hold for this system. However, we can still prove the following weaker result.

THEOREM III.4 (The equivalence rule for the ε*-calculus). *Let X be any set of proper formulae, E and E' any two quasi-terms or quasi-formulae, and A any proper formula or term containing some specified occurrence of E, provided that this occurrence does not lie within the scope of an ε-symbol in A. Let A' be the expression obtained from A by replacing this specified occurrence of E by E'. If A' is a proper formula or term and if $X \vdash_{\varepsilon*} \forall[E \equiv E']$, then $X \vdash_{\varepsilon*} A \equiv A'$*

Proof. The proof is identical to the proof of Theorem II.21 except that we can omit the third part of Case 6, where A is of the form εxB. In this way we avoid using E2-axioms.

THEOREM III.5. *Any proper formula of the form*

$$(s = t \wedge A_s^x) \to A_t^x$$

is a theorem of the ε-calculus, provided that x does not have a free occurrence in A within the scope of an ε-symbol.*

Proof. See the proof of Theorem II.22.

COROLLARY. *Any ε-free formula of the form*

$$(s = t \wedge A_s^x) \to A_t^x$$

is a theorem of the predicate calculus.

3.1 Subordination and rank

In view of Theorem III.3, to complete the proof of the Second ε-Theorem it is sufficient to prove that for any ε-free X and A, if $X \vdash_\varepsilon A$, then $X \vdash_{\varepsilon*} A$. In other words we must prove the eliminability of improper formulae and E2-axioms from deductions of A from X in the ε-calculus, where A and X are ε-free. This is by far the most difficult part of the proof of the Second ε-Theorem, and in order to carry it out we must examine the structure of improper formulae in some detail.

The notion of an improper formula is closely related to Hilbert and Bernays' notion of the subordination of quasi ε-terms. Hilbert and Bernays [1939], page 23, define a quasi ε-term t to be subordinate (*untergeordnet*) to another quasi ε-term εyB if and only if B contains t, and a free occurrence of y in B lies within t. Thus, for example, the quasi ε-term

$$\varepsilon x P x y$$

is subordinate to the term

$$\varepsilon y(y = \varepsilon x P x y).$$

We shall extend this notion of subordination to include subordination within quasi-formulae. If A is a quasi-formula of the form QyB or a quasi-term of the form εyB, then a quasi ε-term t is said to be *subordinate to A* if and only if B contains t, and a free occurrence of y in B lies within t. Clearly, a formula QyB is improper if it contains a quasi ε-term which is subordinate to it. For example, the quasi ε-term

$$\varepsilon xPxy$$

is subordinate to the (improper) formula

$$\forall y(y = \varepsilon xPxy).$$

We shall now assign to every ε-term t a positive integer, called the 'rank' of t. We shall denote this number by $rk(t)$. Intuitively, the rank of t is a measure of the complexity of the subordination within t. Before giving a definition of the rank function, we list the four properties which this function must possess.

Properties of rank:

R1. $rk(t) \geq 1$, and if there are no quasi ε-terms subordinate to t, then $rk(t) = 1$.

R2. If t' is obtained from t by replacing every occurrence of some subterm of t by some other term, then $rk(t') = rk(t)$.

R3. If t is of the form $[t_1]_s^x$ for some term s, and t_1 is subordinate to εxA, then $rk(\varepsilon xA) > rk(t)$.

R4. For any ε-terms εxA and εyB, $rk(\varepsilon z \neg (A_z^x \leftrightarrow B_z^y)) = \max \{rk(\varepsilon xA), rk(\varepsilon yB)\}$.

In what follows, every argument concerning the rank of an ε-term depends only on properties R1–R4, and therefore any definition of the rank function which yields these four properties will suffice. The definition we use is that which is given by Hilbert and Bernays (vol. 2, p. 25).[1] For any quasi ε-term t, $rk(t)$ is defined as follows (by induction on the length of t):

If there are no quasi ε-terms subordinate to t, then $rk(t) = 1$; otherwise $rk(t)$ is one greater than the maximal rank of the quasi ε-terms which are subordinate to t.

For example, $rk(\varepsilon xPxy) = 1$ and $rk(\varepsilon y(y = \varepsilon xPxy)) = 2$.

[1] Property R4 is not needed in proving the Second ε-Theorem for an ε-calculus without axiom schema E2. To prove this weaker theorem the following simpler definition of the rank of t may be used. Let t^0 be the term obtained from t by replacing every subterm of t by the symbol a_1. Then the *rank* of t is defined to be the number of occurrences of the ε-symbol in t^0. This definition clearly satisfies R1–R3.

EXERCISE

Show that the above definition of rank satisfies properties R1–R4.

3.2 r, T_r-deductions

Let $\mathscr{D}$ be any deduction in the ε-calculus. An ε-term εyB is a Q-*term* in $\mathscr{D}$ if the formula

$$\exists yB \rightarrow B(\varepsilon yB)$$

or some formula of the form

$$\neg \exists yB \rightarrow \neg B(\cdot)$$

is used as an axiom in $\mathscr{D}$. The term εyB is an E2-*term* in $\mathscr{D}$ if a formula of the form

$$\forall z(A_z^x \leftrightarrow B_z^y) \rightarrow \varepsilon xA = \varepsilon yB$$

or of the form

$$\forall z(B_z^y \leftrightarrow A_z^x) \rightarrow \varepsilon yB = \varepsilon xA$$

is used as an axiom in $\mathscr{D}$. For any non-negative integer r, a deduction $\mathscr{D}$ in the ε-calculus is an *r-deduction* if every Q-term and every E2-term in $\mathscr{D}$ has rank $\leq r$. Furthermore, if T_r is some finite collection of ε-terms of rank r, then an *r,T_r-deduction* is an r-deduction in which every E2-term of rank r is a member of T_r. Notice that since every ε-term has rank ≥ 1, a 0-deduction is one in which no instances of axiom schemata Q3, Q4, or E2 are used as axioms, and a 1, $\emptyset$-deduction is one in which every Q-term has rank 1 and no instances of axiom schema E2 are used as axioms. To denote that there exists an r-deduction of A from X we write $X \vdash^r A$, and to denote that there exists an r,T_r-deduction of A from X we write $X \vdash^{r,T_r} A$.

It is easy to see that for any r and T_r, the r-deductions and the r,T_r-deductions satisfy the MP rule, tautology rule, f-rule, $\rightarrow$-introduction rule, and $\rightarrow$-elimination rule, since no Q-axioms or E2-axioms are used in verifying these rules. Thus for example

$$X, \neg A \vdash^{r,T_r} f \quad \text{iff} \quad X \vdash^{r,T_r} A,$$

and $$X, A \vdash^{r,T_r} B \quad \text{iff} \quad X \vdash^{r,T_r} A \rightarrow B.$$

In order to prove the ε-Theorems we shall prove the following two results:

(1) For any ε-free X and any $r > 1$, if $X \vdash^{r,0} f$, then $X \vdash^{r-1} f$.
(2) For any ε-free X, any $r \geq 1$, and any finite collection T_r of ε-terms of rank r, if $X \vdash^{r,T_r} f$, then $X \vdash^{r,0} f$

Using these two results (and the f-rule) it then follows that for any ε-free X and ε-free A, if $X \vdash_\varepsilon A$, then $X \vdash^{1,\emptyset} A$.

M.L.—6

The proof of the two results depends on the following theorem concerning the way in which an axiom is affected when an ε-term occurring in that axiom is replaced by some other term.

THEOREM III.6. *Suppose εyB is any ε-term, s any term, C any axiom, and C' the formula obtained from C by replacing each occurrence of εyB by s. Then:*

(i)   *If C is a propositional axiom, Q1-axiom, Q2-axiom, E1-axiom, or E3-axiom, then C' is an axiom of the same form.*

(ii) *If C is an E2-axiom and εyB is neither of the two specified ε-terms in C, then C' is an E2-axiom.*

(iii) *If C is the Q3-axiom $\neg \exists x A \to \neg A(t)$, where $rk(\varepsilon x A) \le rk(\varepsilon y B)$, then C' is a Q3-axiom.*

(iv) *If C is the Q4-axiom $\exists x A \to A(\varepsilon x A)$, where $rk(\varepsilon x A) \le rk(\varepsilon y B)$ and $\varepsilon x A$ is not εyB, then C' is a Q4-axiom.*

Proof. The proofs of (i) and (ii) are trivial. (For a discussion of the E1-axioms see page 50.)

(iii). We shall prove that C' is the Q3-axiom $\neg \exists x A' \to \neg A'(t')$, where A' and t' are obtained from A and t by replacing each occurrence of εyB by s. Suppose this is not the case. Then εyB must be of the form $[p]_t^x$, for some quasi-term p which is subordinate to $\exists x A$. Hence p is subordinate to $\varepsilon x A$, and by rank property R3 we have $rk(\varepsilon y B) < rk(\varepsilon x A)$ which contradicts our assumption that $rk(\varepsilon x A) \le rk(\varepsilon y B)$. In other words our assumption on the ranks of $\varepsilon x A$ and $\varepsilon y B$ rules out the type of situation which we illustrated on page 65, where C was the formula

$$\neg \exists x (x = \varepsilon y(y = x)) \to \neg(t = \varepsilon y(y = t))$$

and εyB the term $\varepsilon y(y = t)$. Note that in this example εyB is $[\varepsilon y(y = x)]_t^x$, and $\varepsilon y(y = x)$ is subordinate to $\exists x(x = \varepsilon y(y = x))$.

(iv). To prove part (iv) we can use the same argument that was employed in (iii) to show that C' must be the Q4-axiom

$$\exists x A' \to A'(\varepsilon x A'),$$

where A' is obtained from A by replacing each occurrence of εyB by s.

3.3   The Rank Reduction Theorem

THEOREM III.7 (The Rank Reduction Theorem). *For any set X of ε-free formulae and any $r > 1$, if $X \vdash^{r,\emptyset} f$, then $X \vdash^{r-1} f$.*

Proof. Let $\mathscr{D}$ be an $r, \emptyset$-deduction of f from X. The proof follows by induction on the number, p, of Q-terms of rank r in $\mathscr{D}$. If $p = 0$, then $\mathscr{D}$ itself is an $(r - 1)$-deduction of f from X. Suppose $p > 0$. Let εyB be some Q-term of rank r in $\mathscr{D}$ whose length is at least as great as that of every other Q-term of rank r in $\mathscr{D}$. Let S be the collection of remaining $p-1$ Q-terms of rank r. Thus εyB is not contained in any member of S. For the purposes of this proof we write $Y \vdash' A$ to denote that there exists an $r, \emptyset$-deduction of A from Y in which every Q-term of rank r is a member of S. It is sufficient to prove $X \vdash' f$, since we may then apply the induction hypothesis.

Suppose that $\mathscr{D}$ is the sequence $\langle A_1, \ldots, A_m \rangle$. For each $i = 1, \ldots, m$, let A_i' be the formula obtained from A_i by replacing each occurrence of $\exists yB$ by $B(\varepsilon yB)$ and, if B is of the form $\neg C$, by replacing each occurrence of $\forall yC$ by $C(\varepsilon y \neg C)$, i.e., $C(\varepsilon yB)$. Thus if A_i is the Q4-axiom $\exists yB \to B(\varepsilon yB)$, then A_i' is the tautology

$$(1) \qquad B(\varepsilon yB) \to B(\varepsilon yB).$$

If B is of the form $\neg C$ and A_i is the Q1-axiom $\forall yC \to \neg \exists y \neg C$, then A_i' is the tautology

$$(2) \qquad C(\varepsilon y \neg C) \to \neg \neg C(\varepsilon y \neg C),$$

and if A_i is the Q2-axiom $\neg \forall yC \to \exists y \neg C$, then A_i' is the tautology

$$(3) \qquad \neg C(\varepsilon y \neg C) \to \neg C(\varepsilon y \neg C).$$

Finally, if A_i is the Q3-axiom $\neg \exists yB \to \neg B(t)$, for some term t, then A_i' is the formula

$$(4) \qquad \neg B(\varepsilon yB) \to \neg B(t).$$

If A_i is any other axiom, then A_i' is an axiom of the same form. (Using rank property R4 and the fact that every E2-term in $\mathscr{D}$ has rank $< r$, verify that if A_i is an E2-axiom, then so is A_i'.) Finally if A_i is a member of X, then A_i' is A_i. For, since $rk(\varepsilon yB) > 1$, then $\exists yB$ (and $\forall yC$) are not ε-free. Therefore, since every member of X is ε-free, the members of X are unaffected.

Consequently, if we augment the sequence $\langle A_1', \ldots, A_m' \rangle$ by including proofs of the tautologies (1), (2), and (3), and if we regard the formulae of the form (4) as assumptions, we obtain a deduction of f from

$$X \cup \{\neg B(\varepsilon yB) \to \neg B(t_1), \ldots, \neg B(\varepsilon yB) \to \neg B(t_n)\}$$

for some terms $t_1, \ldots, t_n$. Furthermore by rank property R2 and our choice of εyB with maximal length this deduction is an $r, \emptyset$-deduction such that every Q-term of rank r is still a member of S. Hence

$$(5) \qquad X, \neg B(\varepsilon yB) \to \neg B(t_1), \ldots, \neg B(\varepsilon yB) \to \neg B(t_n) \vdash' f.$$

Now using the tautologies

$$B(\varepsilon yB) \to \neg B(\varepsilon yB) \to \neg B(t_i)$$

and Theorem II.3(ii), (5) implies

(6) $X, B(\varepsilon y B) \vdash' f,$

and therefore

(7) $X \vdash' \neg B(\varepsilon y B)$

by the f-rule.

Now let $\langle C_1, \ldots, C_k \rangle$ be this r, $\emptyset$-deduction of $\neg B(\varepsilon y B)$ from X. For each C_i and each term t_j let C_i^j be the formula obtained from C_i by replacing each occurrence of $\varepsilon y B$ by t_j. Then for each $j = 1, \ldots, n$, the sequence $\langle C_1^j, \ldots, C_k^j \rangle$ is an r, $\emptyset$-deduction of $\neg B(t_j)$ from X, by Theorem III.6. Furthermore by our choice of $\varepsilon y B$ with maximal length, every Q-term of rank r in this deduction is still a member of S. Hence, for each $j = 1, \ldots, n$

(8) $X \vdash' \neg B(t_j),$

and consequently by the tautology rule

(9) $X \vdash' \neg B(\varepsilon y B) \rightarrow \neg B(t_j).$

The desired result, $X \vdash' f$, now follows from (5) and (9) by Theorem II.3(ii). This completes the proof.

Notice that the above proof depends on two essential conditions: (i) that $r > 1$, and (ii) that X is ε-free. We impose these conditions in order to ensure that the replacements of $\exists y B$ by $B(\varepsilon y B)$, $\forall y C$ by $C(\varepsilon y \neg C)$, and $\varepsilon y B$ by t_j do not affect any members of X. In the following theorem we consider the special cases where $r = 1$ and where X is a certain set of formulae which may contain the ε-symbol. This theorem will be used later as the central lemma in proving the First ε-Theorem.

Recall that a quasi-formula is elementary if it contains no occurrences of the symbols $\forall$, $\exists$, or ε.

THEOREM III.8. *Let Y be any set of elementary quasi-formulae and let X be any set of substitution instances of members of Y. (The members of X are not necessarily ε-free.) If $X \vdash^{1,\emptyset} f$, then there exists a set Z of substitution instances of members of Y such that $Z \vdash_{EC} f$.*

Proof. The proof is identical to that of the Rank Reduction Theorem except that we may no longer assume that the various replacement procedures used in this proof leave the members of X unaffected. However, since any member of X is a formula of the form

$$B^{x_1 \, \cdots \, x_n}_{t_1 \, \cdots \, t_n}$$

where B is an elementary quasi-formula, then any member of X which is affected is converted into another substitution instance of the same quasi-formula. Consequently, the proof of the Rank Reduction Theorem yields a 0-deduction $\mathscr{D}$ of f from some set X' of substitution instances of members of

Y. Now a 0-deduction has no Q-terms or E2-terms. Consequently no Q3, Q4, or E2 axioms are used in $\mathscr{D}$. If throughout $\mathscr{D}$ we replace every quasi-formula of the form $\forall y C$ by f and every quasi-formula of the form $\exists x B$ by $\neg f$ (starting with those of maximal length), the Q1-axioms and Q2-axioms are converted into tautologies and the other axioms are converted into new axioms of the same form. If we then replace every remaining ε-term by some individual symbol a, we obtain a deduction of f from a set Z of substitution instances of members of Y. Since every formula in this deduction is elementary, then $Z \vdash_{EC} f$.

3.4 The E2 Elimination Theorem

To complete the proof of the Second ε-Theorem we need to show that the E2-axioms are eliminable. This is the essence of the following theorem.

THEOREM III.9 (The E2 Elimination Theorem). *For any ε-free X, any $r \geq 1$, and any finite collection T_r of ε-terms of rank r, if $X \vdash^{r,T_r} f$, then $X \vdash^{r,0} f$.*

Proof. The proof is by induction on the number n of terms in T_r. If $n = 0$, there is nothing to prove. Suppose $n > 0$. Let $\varepsilon x A$ be a member of T_r whose length is at least as great as that of every other term in T_r. Let S_r be the collection of remaining $n - 1$ terms in T_r. By our choice of $\varepsilon x A$ with maximal length it follows that $\varepsilon x A$ does not occur in any member of S_r. To prove the theorem it is sufficient to prove $X \vdash^{r,S_r} f$, since we may then apply the induction hypothesis.

Let $\mathscr{D}$ be any r, T_r-deduction of f from X. We may assume that $\varepsilon x A$ is an E2-term in $\mathscr{D}$, since otherwise there would be nothing to prove. Secondly, we may assume that no E2-axiom of the form

$$\forall z(A_z^x \leftrightarrow A_z^x) \to \varepsilon x A = \varepsilon x A$$

is used as an axiom in $\mathscr{D}$, since any formula of this form follows by modus ponens from axiom schema P1 and axiom schema E3. Finally, we may assume that every E2-axiom, involving $\varepsilon x A$, which is used in $\mathscr{D}$ is of the form

$$\forall z(A_z^x \leftrightarrow B_z^x) \to \varepsilon x A = \varepsilon y B,$$

i.e., $\varepsilon x A$ is on the left-hand side of the identity symbol. This assumption can easily be justified by observing that

$$\vdash^{r,0} \forall z(B_z^y \leftrightarrow A_z^x) \to \forall z(A_z^x \leftrightarrow B_z^y),$$

and
$$\vdash^{1,0} \varepsilon x A = \varepsilon y B \to \varepsilon y B = \varepsilon x A.$$

Let $E_1, \ldots, E_m$ be the E2-axioms, involving $\varepsilon x A$, which are used in $\mathscr{D}$. If these are regarded as assumptions, then $\mathscr{D}$ is an r, S_r-deduction of f from $X \cup \{E_1, \ldots, E_m\}$. Thus

$$X, E_1, \ldots, E_m \vdash^{r,S_r} f.$$

Consequently, to prove $X \vdash^{r,S_r} f$, it is sufficient by virtue of Theorem II.3(ii) to prove for each $i = 1, \ldots, m$

(1) $$X \vdash^{r,S_r} E_i.$$

Take any E_i. Let it be the formula

E_i $$\forall z(A_z^x \leftrightarrow B_z^y) \to \varepsilon x A = \varepsilon y B.$$

Thus either (i) $rk(\varepsilon y B) < r$ or (ii) $rk(\varepsilon y B) = r$ and $\varepsilon y B$ is a member of S_r. By the tautology rule and the f-rule, to prove (1) it is sufficient to prove

(2) $$X, \forall z(A_z^x \leftrightarrow B_z^y) \vdash^{r,S_r} f.$$

The proof of (2) is as follows. Throughout the original deduction $\mathscr{D}$ of f from X replace every occurrence of $\varepsilon x A$ by $\varepsilon y B$. For any formula A_j which is used as an axiom in $\mathscr{D}$, if A_j is not one of the E2-axioms $E_1, \ldots, E_m$ and not the Q3-axiom $\exists x A \to A(\varepsilon x A)$, then by Theorem III.6, the formula obtained from A_j by replacing $\varepsilon x A$ by $\varepsilon y B$ is an axiom of the same form. Furthermore since X is ε-free, the members of X are unaffected by the replacement of $\varepsilon x A$ by $\varepsilon y B$. Consequently, it follows that

(3) $$X, \exists x A \to A(\varepsilon y B), E_1', \ldots, E_m' \vdash^{r,S_r} f,$$

where for each $j = 1, \ldots, m$, E_j' is the formula obtained from E_j by replacing each occurrence of $\varepsilon x A$ by $\varepsilon y B$. (Why is the deduction still an r, S_r-deduction?)

Now to prove (2) it is sufficient by virtue of (3) and Theorem II.3(ii) to prove that

(4) $$\forall z(A_z^x \leftrightarrow B_z^y) \vdash^{r,\emptyset} \exists x A \to A(\varepsilon y B),$$

and for each $j = 1, \ldots, m$,

(5) $$\forall z(A_z^x \leftrightarrow B_z^y) \vdash^{r,S_r} E_j'.$$

Proof of (4): Let Y be $\{\forall z(A_z^x \leftrightarrow B_z^y)\}$. Since $\varepsilon x A$ and $\varepsilon y B$ have rank $\leq r$, the term $\varepsilon z \neg(A_z^x \leftrightarrow B_z^y)$ also has rank $\leq r$ by rank property R4. Therefore these three terms may be used as Q-terms in the desired r, $\emptyset$-deduction of $\exists x A \to A(\varepsilon y B)$ from Y. This deduction can be constructed as follows:

$Y \vdash^{r,\emptyset} \forall z(A_z^x \leftrightarrow B_z^y)$	
$Y \vdash^{r,\emptyset} A(\varepsilon x A) \leftrightarrow B(\varepsilon x A)$	∀-elimination
$Y \vdash^{r,\emptyset} A(\varepsilon y B) \leftrightarrow B(\varepsilon y B)$	∀-elimination
$Y \vdash^{r,\emptyset} \exists x A \to A(\varepsilon x A)$	Q4-axiom
$Y \vdash^{r,\emptyset} \neg \exists y B \to \neg B(\varepsilon x A)$	Q3-axiom
$Y \vdash^{r,\emptyset} \exists y B \to B(\varepsilon y B)$	Q4-axiom
$Y \vdash^{r,\emptyset} \exists x A \to A(\varepsilon y B)$	tautology rule

Proof of (5): Suppose E_j is the formula

$$\forall u(A_u^x \leftrightarrow C_u^v) \to \varepsilon x A = \varepsilon v C.$$

Then E_j' is the formula

$$\forall u(A_u^x \leftrightarrow C'\,_u^v) \to \varepsilon y B = \varepsilon v C',$$

where C' is obtained from C be replacing every occurrence of $\varepsilon x A$ (if any) by $\varepsilon y B$. Recall that each of the terms $\varepsilon y B$ and $\varepsilon v C$ satisfies the condition that either (i) its rank is less than r or (ii) its rank equals r and it is a member of S_r. Consequently by rank property R2 and the fact that $\varepsilon x A$ does not occur in any member of S_r, the term $\varepsilon v C'$ also satisfies this condition. Therefore the terms $\varepsilon y B$ and $\varepsilon v C'$ may be used as E2-terms in the desired r, S_r-deduction of E_j' from $\forall z(A_z^x \leftrightarrow B_z^y)$. This deduction can be constructed as follows. Let Y be the set

$$\{\forall z(A_z^x \leftrightarrow B_z^y), \forall u(A_u^x \leftrightarrow C'\,_u^v)\}$$

and let t be the term $\varepsilon w \neg(B_w^y \leftrightarrow C'\,_w^v)$, where w is some variable which is free for y in B and free for v in C'. Then

$$
\begin{array}{lll}
Y \vdash^{r,S_r} \forall z(A_z^x \leftrightarrow B_z^y) & & \\
Y \vdash^{r,S_r} \forall u(A_u^x \leftrightarrow C'\,_u^v) & & \\
Y \vdash^{r,S_r} [A_z^x \leftrightarrow B_z^y]_t^z & & \forall\text{-elimination} \\
Y \vdash^{r,S_r} [A_u^x \leftrightarrow C'\,_u^v]_t^u & & \forall\text{-elimination} \\
Y \vdash^{r,S_r} [B_w^y \leftrightarrow C'\,_w^v]_t^w & & \text{tautology rule} \\
Y \vdash^{r,S_r} \forall w(B_w^y \leftrightarrow C'\,_w^v) & & \forall\text{-introduction} \\
Y \vdash^{r,S_r} \forall w(B_w^y \leftrightarrow C'\,_w^v) \to \varepsilon y B = \varepsilon v C' & & \text{E2-axiom} \\
Y \vdash^{r,S_r} \varepsilon y B = \varepsilon v C' & & \text{modus ponens} \\
\forall z(A_z^x \leftrightarrow B_z^y) \vdash^{r,S_r} E_j' & & \to\text{-introduction}
\end{array}
$$

This completes the proof of Theorem III.9.

3.5   The First and Second ε-Theorems

We have now established all the major results which are needed in proving the Second ε-Theorem. However, in order to prove the First ε-Theorem we still need the following simple lemma.

LEMMA. *For any prenex formula A, there exists a substitution instance A' of the matrix of A such that $\vdash^{1,\emptyset} A' \leftrightarrow A$.*

Proof. The proof depends on the following fact. If $rk(\varepsilon y B) = 1$, then

(1) $$\vdash^{1,\emptyset} B(\varepsilon y B) \leftrightarrow \exists y B, \quad \text{and}$$

(2) $$\vdash^{1,\emptyset} B(\varepsilon y \neg B) \leftrightarrow \forall y B.$$

This fact is an obvious consequence of Theorem II.10. Now let A be the prenex formula

$$Q_1 x_1 \ldots Q_n x_n B,$$

where B is its matrix. For each $i = 1, \ldots, n$, we define the term t_i as follows:

$$t_1 \text{ is } \begin{cases} \varepsilon x_1 Q x_2 \ldots Q x_n B & \text{if } Q_1 \text{ is } \exists \\ \varepsilon x_1 \neg Q x_2 \ldots Q x_n L & \text{if } Q_1 \text{ is } \forall, \end{cases}$$

$$t_2 \text{ is } \begin{cases} \varepsilon x_2 Q x_3 \ldots Q x_n [B]_{t_1}^{x_1} & \text{if } Q_2 \text{ is } \exists \\ \varepsilon x_2 \neg Q x_3 \ldots Q x_n [B]_{t_1}^{x_1} & \text{if } Q_2 \text{ is } \forall, \end{cases}$$

and so on. Now let A' be the substitution instance $B_{t_1 \ldots t_n}^{x_1 \ldots x_n}$ of B. Since B is ε-free (by the definition of a prenex formula), then $rk(t_i) = 1$, for each i. Consequently, using induction on n we can prove $\vdash A' \leftrightarrow A$ from (1) and (2).

THEOREM III.10 (The First ε-Theorem). *If X is any set of prenex formulae, A is any prenex formula, and $X \vdash_{PC} A$, then $Z \vdash_{EC} B_1 \vee \ldots \vee B_n$, where each member of Z is some substitution instance of the matrix of some member of X and each B_i is some substitution instance of the matrix of A.*

Proof. We shall first prove the theorem for the special case where A is the formula f.

Assume $X \vdash_{PC} f$. Then there exists a finite subset X_1 of X such that $X_1 \vdash_{PC} f$. Let Y be the (finite) set consisting of the matrices of the members of X_1. By Theorem III.2 there exists a deduction of f from X_1 in the ε*-calculus, and therefore $X_1 \vdash^{1,0} f$. Now let X_2 be the finite set of substitution instances of members of Y which correspond to each member of X_1 according to the above lemma. Since $X_2 \vdash^{1,0} C$ for each C in X_1, then by Theorem II.3(ii), $X_2 \vdash^{1,0} f$. Therefore by Theorem III.8 there exists a set Z of substitution instances of members of Y such that $Z \vdash_{EC} f$. This proves the theorem for the special case where A is the formula f.

Now suppose that A is any prenex formula of the form $Q_1 x_1 \ldots Q_k x_k C$, where C is its matrix. Let B be the prenex formula $Q_1' x_1 \ldots Q_k' x_k \neg C$, where as usual Q_i' is $\exists$ if Q_i is $\forall$, and Q_i' is $\forall$ if Q_i is $\exists$. Note that any substitution instance of B is the negation of the corresponding substitution instance of A. Since $\vdash_{PC} A \leftrightarrow \neg B$, then $X \vdash_{PC} A$ implies $X, B \vdash_{PC} f$. Therefore by the above special case there exists a set Z such that $Z \vdash_{EC} f$ and every member of Z is a substitution instance of the matrix of some member of X or a substitution instance of $\neg C$. Let $\neg B_1, \ldots, \neg B_n$ be the substitution instances of $\neg C$ and let $A_1, \ldots, A_m$ be the remaining members of Z. Since

$$\neg B_1, \ldots, \neg B_n, A_1, \ldots, A_m \vdash_{EC} f,$$

repeated applications of the α-rule (Theorem II.12) yield

$$\neg(B_1 \vee \ldots \vee B_n), A_1, \ldots, A_m \vdash_{EC} f,$$

and the desired result now follows by the f-rule.

THEOREM III.11 (The Second ε-Theorem). *For any ε-free X and any ε-free A, if $X \vdash_\varepsilon A$, then $X \vdash_{PC} A$.*

Proof. Since $X \vdash_\varepsilon A$, then $X, \neg A \vdash_\varepsilon f$ by the f-rule. By the Rank Reduction Theorem and the E2-Elimination Theorem this implies that there exists a $1, \emptyset$-deduction of f from $X \cup \{\neg A\}$, and therefore a $1, \emptyset$-deduction $\mathscr{D}$ of A from X. We want to convert the deduction $\mathscr{D}$ into a deduction of A from X in the ε*-calculus, since we can then apply Theorem III.3 to get a deduction of A from X in the predicate calculus. The only difficulty is that the formulae in $\mathscr{D}$ are not necessarily proper. Consequently, throughout $\mathscr{D}$ we replace every quasi ε-term which contains a free variable by the symbol a (starting with those quasi-terms of maximal length.) Since every Q-term in $\mathscr{D}$ has rank 1, it is easy to see that this procedure does not damage any of the axioms and therefore provides a deduction of A from X in the ε*-calculus.

4.1 Skolem's Theorem

One of the most important consequences of the Second ε-Theorem is the counterpart of Theorem II.28 for the predicate calculus. This result is commonly known as Skolem's Theorem.

THEOREM III.12 (Skolem's Theorem). *Let A be any prenex formula of $\mathscr{L}(\mathscr{V})$, C any ε-free formula of $\mathscr{L}(\mathscr{V})$, X any set of ε-free formulae of $\mathscr{L}(\mathscr{V})$, and $\mathscr{V}^*$ the vocabulary obtained from $\mathscr{V}$ by adjoining the Skolem functions used in forming the Skolem resolution A_S of A. Then*

$$X, A \vdash_{PC(\mathscr{V})} C \quad iff \quad X, A_S \vdash_{PC(\mathscr{V}^*)} C.$$

Proof. Use Theorem II.28 and the Second ε-Theorem. (*Note*: The half of the theorem which states that $X, A \vdash_{PC(\mathscr{V})} C$ implies $X, A_S \vdash_{PC(\mathscr{V}^*)} C$ is much 'weaker' than the other half and can be proved directly using only the axioms and rules of inference of the predicate calculus.)

Notice that Theorems II.25, II.26, and II.27 also hold for the predicate calculus by virtue of the Second ε-Theorem. The analogue of Theorem II.26 now provides us with a full justification (in terms of the predicate calculus) of the familiar 'rule of inference' used by mathematicians when they pass from the statement

'for all x, there exists a y such that $B(x,y)$'

to the statement

'for all x, $B(x,g(x))$',

or, to put it more colloquially, to the statement

'for each x, let $g(x)$ be some y such that $B(x,y)$',

provided that the function g does not appear in the statement $B(x,y)$, in the statement being proved, or in any of the assumptions on which the proof is based (cf. Chapter II, §9). We can state the formal justification of this rule as follows.

THEOREM III.13. *If* $X \vdash_{PC} \forall x_1 \ldots \forall x_n \exists y B$ *and* $Y, \forall x_1 \ldots \forall x_n [B]^y_{gx_1 \ldots x_n} \vdash_{PC} C$, *then* $X \cup Y \vdash_{PC} C$, *provided that g does not occur in C, B, or any member of Y.*

Proof. By Theorem III.2, Theorem II.26, and the Second ε-Theorem we have $Y, \forall x_1 \ldots \forall x_n \exists y B \vdash_{PC} C$. Hence $X \cup Y \vdash_{PC} C$ by Theorem II.32.

Unfortunately, the restriction that g does not occur in any member of Y limits the applicability of this rule in ordinary mathematical arguments. In most branches of mathematics the axioms of set theory are used as basic assumptions. These axioms usually include certain formulae known as *axioms of replacement* (cf. p. 106). In order to apply the above rule concerning the eliminability of the new function symbol g, one must make sure that the set of assumptions Y does not include an axiom of replacement in which the symbol g occurs. For example, if one uses the fact that the values of $g(x)$ form a set when x ranges over some set, then one is using an axiom of replacement which contains g. In this case the above rule can only be justified if the axiom of choice is included as one of the basic assumptions.

However, one does not need to appeal to the axiom of choice if one has proved the stronger statement

'for all x, there exists a *unique* y such that $B(x,y)$'.

In this case the function symbol g can be eliminated even if it is used in an axiom of replacement. The proof of this fact does not depend on the Second ε-Theorem, but rather on the eliminability of the ι-symbol (cf. p. 100).

We shall return to these problems in Chapter IV when we consider formalizations of set theory based on the ε-calculus.

4.2   Herbrand's Theorem

Herbrand's Theorem, as it was originally formulated (Herbrand [1930]), involves the complicated notions of 'properties B and C of order p'. In order to simplify both the statement and proof of this theorem various people have proved results which bear the title 'Herbrand's Theorem', but which are considerably weaker than Herbrand's original assertion. This situation, together with the relative inaccessibility of Herbrand's paper, has doubtless given rise to a certain amount of confusion about the exact nature of this theorem. In this section we shall give a relatively simple formulation and proof of Herbrand's Theorem which incorporates most of the essential features of the original assertion.

The essence of Herbrand's Theorem can be described as follows. Let A be any identity-free, prenex formula. For each positive integer p we assign to A a particular elementary formula, called the 'p-reduction' of A. We say that A is a 'p-tautology' if the p-reduction of A is a tautology. The theorem then states that A is a theorem of the predicate calculus if and only if A is a p-tautology for some $p > 0$. This theorem is a very powerful one since it establishes a useful necessary and sufficient condition for provability in the predicate calculus. One of its main applications has been in solving various cases of the decision problem for the predicate calculus.

Before defining the p-reduction of A, we need the following definitions. For any term t, the *degree* of t is defined as follows by induction on its length:

1. If t has no subterms, the degree of t is 1.
2. If t has subterms, the degree of t is one greater than the maximal degree of all its subterms.

For example, the degree of any individual symbol or 0-place function symbol is 1, and the degree of the term $g^2 a h^1 a$ is 3.

Let $\mathscr{G}$ be any finite collection of individual symbols and function symbols. Then a $\mathscr{G}$-*term* is any term whose symbols are all members of $\mathscr{G}$.

Now let A be any prenex formula, let A_H be its Herbrand resolution, let B be the matrix of A_H, and let $\mathscr{G}$ be the collection of individual symbols and function symbols occurring in B. (If B contains no individual or 0-place function symbols, we take a_1 as an additional member of $\mathscr{G}$.) For any positive integer p, a p-*substitution instance* of B, is any substitution instance

$$B^{x_1 \ldots x_n}_{t_1 \ldots t_n}$$

of B where each t_i is a $\mathscr{G}$-term with degree $\leq p$. Thus for any p, there are finitely many p-substitution instances of B. The p-*reduction* of A is now defined as the disjunction, $B_1 \vee \ldots \vee B_m$, of all the p-substitution instances of B. We say that A is a p-*tautology* if its p-reduction is a tautology. (To be exact, the p-reduction of A depends not only on A and p, but also on the order of its disjunctive parts, $B_1, \ldots, B_m$, and on the particular choice of Herbrand functions used in forming A_H. However, for any two p-reductions of A, the one is a tautology if and only if the other is. Consequently, the notion of a p-tautology is well-defined.)

Clearly, if A is a p-tautology, then for any $q \geq p$, A is also a q-tautology. For, if $q \geq p$, then every p-substitution instance of the matrix of A_H is also a q-substitution instance, and therefore the p-reduction of A is a 'sub-disjunction' of the q-reduction of A.

To illustrate the above definitions, consider the following example. Let A be the formula

$$\exists y \forall z (Py \rightarrow Pz),$$

where P is some 1-place predicate symbol. Then the Herbrand resolution, A_H, of A is

$$\exists y (Py \rightarrow Pgy),$$

where g is a 1-place function symbol. Thus we let $\mathscr{G}$ be the set $\{a_1, g\}$. The only $\mathscr{G}$-term of degree 1 is a_1, the only $\mathscr{G}$-term of degree 2 is ga_1, etc. Consequently, the 1-reduction of A is

(1) $Pa_1 \rightarrow Pga_1$

and the 2-reduction of A is

(2) $(Pa_1 \rightarrow Pga_1) \vee (Pga_1 \rightarrow Pgga_1).$

Although (1) is not a tautology, (2) is a tautology. Hence A is a p-tautology for any $p \geqslant 2$.

We can now give the statement and proof of Herbrand's Theorem.

THEOREM III.14 (Herbrand's Theorem). *Let A be any identity-free prenex formula. Then*:
 (i)   *If $\vdash_{PC} A$, then there exists a positive integer p such that A is a p-tautology.*
(ii)   *If A is a p-tautology for some positive integer p, then $\vdash_{PC'} A$.*

Proof. (i) Assume $\vdash_{PC} A$. Let A_H be the Herbrand resolution of A, B the matrix of A_H, and $\mathscr{G}$ the collection of individual symbols and function symbols occurring in B. ($\mathscr{G}$ also contains a_1 if necessary.) Since $\vdash_{PC} A$, then $\vdash_{\varepsilon} A$ by Theorem III.2, and therefore $\vdash_{\varepsilon} A_H$ by Theorem II.27. By the First and Second ε-Theorems this implies that there exist substitution instances $B_1, \ldots, B_n$ of B such that $\vdash_{EC} B_1 \vee \ldots \vee B_n$. Since A is identity-free, this implies by Theorem II.30 that $B_1 \vee \ldots \vee B_n$ is a tautology. If any of the substituted terms in $B_1 \vee \ldots \vee B_n$ are not $\mathscr{G}$-terms, replace every occurrence of such terms by some $\mathscr{G}$-term of degree 1. In this way we obtain substitution instances $B_1', \ldots, B_n'$ of B such that $B_1' \vee \ldots \vee B_n'$ is still a tautology. Let p be the maximal degree of all the substituted terms in $B_1' \vee \ldots \vee B_n'$. (If there are no such terms let $p = 1$.) Since each B_i' is a p-substitution instance of B, then the tautology $B_1' \vee \ldots \vee B_n'$ is a sub-disjunction of the p-reduction of A. Hence A is a p-tautology.

 (ii) Conversely, assume that A is a p-tautology for some $p > 0$. The knowledge of p enables us to form the p-reduction of A, which we shall denote by $B_1 \vee \ldots \vee B_n$. Since $B_1 \vee \ldots \vee B_n$ is a tautology, then by Theorem II.29, $\vdash_{EC} B_1 \vee \ldots \vee B_n$ and therefore $\vdash_{PC'} B_1 \vee \ldots \vee B_n$. However, since each B_i is a substitution instance of the matrix of the $\exists$-prenex formula

A_H, then $B_i \vdash_{PC'} A_H$ for each $i = 1, \ldots, n$. Consequently, by repeated applications of the β-rule, we have $B_1 \vee \ldots \vee B_n \vdash_{PC'} A_H$, and therefore $\vdash_{PC'} A_H$. By Theorem III.2, this implies that there exists a proof of A_H in the ε-calculus in which every formula is identity-free. Consequently, by Theorem II.27 and the Second ε-Theorem we have $\vdash_{PC'} A$. This completes the proof of Herbrand's Theorem.

It should be mentioned that in Herbrand's original statement of the theorem A is an arbitrary formula rather than a prenex formula. Since any ε-free formula is equivalent to some prenex formula, our version of the theorem is no weaker than Herbrand's, except that Herbrand attempts to reveal the relationship between an arbitrary formula and the various prenex equivalents of that formula. However, Dreben, Andrews, and Aanderaa [1963] have discovered errors in Herbrand's proof. A corrected proof has been produced by Denton and Dreben [1969].

EXERCISES

1. Using Herbrand's Theorem, prove that the formula $\forall x \exists y Pxy \rightarrow \exists y \forall x Pxy$ is not a theorem of the predicate calculus (See the exercise on page 54.)

2. Let A be any identity-free, prenex formula of the form

$$\forall x_1 \ldots \forall x_m \exists y_1 \ldots \exists y_n B,$$

where B is the matrix of A and no n-place function symbols occur in B for $n > 0$. Describe a decision procedure for determining whether or not A is a theorem of the predicate calculus without identity.

4.3　The eliminability of the identity symbol

Herbrand's Theorem can be used to prove that the identity symbol can be eliminated from proofs in the predicate calculus of formulae which do not themselves contain the identity symbol.

THEOREM III.15. *If A is an identity-free formula and $\vdash_{PC} A$, then $\vdash_{PC'} A$.*

Proof. We may assume that A is a prenex formula since for any ε-free, identity-free formula A there exists an identity-free prenex formula A' such that $\vdash_{PC'} A \leftrightarrow A'$ (cf. the proof of Theorem II.24). Since $\vdash_{PC} A$ and A is identity-free, then by the first half of Herbrand's Theorem there exists a positive integer p such that A is a p-tautology. By the second half of Herbrand's Theorem this implies $\vdash_{PC'} A$.

The eliminability of the $=$-symbol does not hold for the ε-calculus since the E2-axioms are needed for proving theorems which do not themselves contain the identity-symbol. For example, if P, Q, and R are 1-place predicate symbols, the formula

$$(\forall z(Pz \leftrightarrow Qz) \wedge R\varepsilon xPx) \rightarrow R\varepsilon yQy$$

is a theorem of the ε-calculus by virtue of axioms E1 and E2, but there exists no identity-free proof of this formula.

EXERCISE

Modify the axioms of the ε-calculus to obtain an equivalent formal system in which the identity symbol is eliminable. (*Hint*: see Chapter I, §6.)

CHAPTER IV

FORMAL THEORIES

1 Introduction

The objective of this chapter is to reveal the role which the ε-symbol can play in the study of formal theories. First of all, we shall explain how the formalists used the ε-symbol and the ε-Theorems in constructing finitary consistency proofs of various formal theories. Secondly, we shall see how the formulation of certain theories can be simplified if the ε-calculus rather than the predicate calculus is used as the underlying formal system. Lastly, we shall investigate the relationship between the ε-symbol and the axiom of choice in formal set theory.

A *formal theory* $\mathcal{T}$ consists of a vocabulary $\mathcal{V}$, a formal system $\mathcal{F}$ for $\mathcal{V}$, and a set $\mathcal{A}$ of formulae of $\mathcal{L}(\mathcal{V})$. Thus we may regard a formal theory $\mathcal{T}$ as an ordered triple $\langle \mathcal{V}, \mathcal{F}, \mathcal{A} \rangle$. The members of $\mathcal{A}$ are called the (*non-logical*) *axioms of $\mathcal{T}$*, $\mathcal{V}$ is called the *vocabulary of $\mathcal{T}$*, and $\mathcal{F}$ is called the *underlying formal system of $\mathcal{T}$*. Alternatively, we often say that $\mathcal{T}$ is *based on $\mathcal{F}$*. By an abuse of language we often say a 'formula of $\mathcal{T}$' instead of a 'formula of $\mathcal{L}(\mathcal{V})$'. Furthermore, if $\mathcal{T}$ is based on the predicate calculus we say a 'formula of $\mathcal{T}$' instead of an '*ε-free* formula of $\mathcal{L}(\mathcal{V})$'. Throughout this chapter most of the formal theories we deal with are based either on the predicate calculus or on the ε-calculus.

Let $\mathcal{T}$ be some formal theory $\langle \mathcal{V}, \mathcal{F}, \mathcal{A} \rangle$. A formula A of $\mathcal{L}(\mathcal{V})$ is a *theorem* of $\mathcal{T}$ if and only if $\mathcal{A} \vdash_{\mathcal{F}} A$. The theory $\mathcal{T}$ is *inconsistent* if f is a theorem of $\mathcal{T}$; otherwise it is *consistent*. A theory $\mathcal{T}'$ is an *extension* of a theory $\mathcal{T}$ is every formula of $\mathcal{T}$ is a formula of $\mathcal{T}'$ and every theorem of $\mathcal{T}$ is a theorem of $\mathcal{T}'$. Two theories are said to be *equivalent* if each is an extension of the other. An extension $\mathcal{T}'$ of $\mathcal{T}$ is an *inessential extension* of $\mathcal{T}$ if every formula of $\mathcal{T}$ which is a theorem of $\mathcal{T}'$ is also a theorem of $\mathcal{T}$. Obviously if $\mathcal{T}'$ is an inessential extension of $\mathcal{T}$, then $\mathcal{T}$ is consistent if and only if $\mathcal{T}'$ is consistent.

2 Finitary consistency proofs

Suppose $\mathcal{T}$ is a theory based on the predicate calculus. If there exists a model $\mathfrak{M}$ which satisfies the set $\mathcal{A}$ of axioms of $\mathcal{T}$, it follows that $\mathcal{T}$ must be consistent. For, the soundness of the predicate calculus implies that every theorem of $\mathcal{T}$ must be true in the model $\mathfrak{M}$. Since the formula f is false in all models, then f is not a theorem of $\mathscr{T}$, and $\mathscr{T}$ is consistent. The objection to this type of consistency proof is that a very strong metatheory is required. For example, the cardinality of the model may be infinite, and in order to show that f is not a theorem of $\mathscr{T}$ one must use non-constructive arguments to show that the axioms of the predicate calculus are true in this model. One of the main contributions of the formalists was in showing that for certain special theories this type of consistency proof can be carried out in a completely finitary way by appealing to the First ε-Theorem rather than to the soundness of the predicate calculus.

To illustrate the formalists' method we shall prove the consistency of the following simple theory S.

Let $\mathscr{V}$ be the vocabulary consisting of the single 2-place predicate symbol $<$, where as usual we write $s < t$ instead of $<st$, and let $\mathscr{A}$ be the set consisting of the following five formulae:

S1	$\forall x \neg (x < x),$
S2	$\forall x \forall y \forall z((x < y \wedge y < z) \rightarrow x < z),$
S3	$\forall x \forall y(x < y \vee y < x \vee y = x),$
S4	$\forall x \exists y(x < y),$
S5	$\exists x \forall y(x = y \vee x < y).$

We define S to be the formal theory $\langle \mathscr{V}, \mathrm{PC}(\mathscr{V}), \mathscr{A} \rangle$. Obviously, the model consisting of the set of positive integers with the usual ordering relation satisfies the set $\mathscr{A}$, and therefore by a non-finitary argument it follows that S is consistent. However, despite the fact that every model which satisfies $\mathscr{A}$ has infinite cardinality we can still prove the consistency of S in a completely finitary way.

First of all, we modify S so that its axioms are all $\forall$-prenex formulae. Let $\mathscr{V}'$ be the vocabulary obtained from $\mathscr{V}$ by adjoining the 0-place function symbol c and the 1-place function symbol g, and let $\mathscr{A}'$ be the set of formulae obtained from $\mathscr{A}$ by replacing formulae S4 and S5 by

S4'	$\forall x(x < gx),$
S5'	$\forall y(c = y \vee c < y).$

We define S' to be the formal theory $\langle \mathscr{V}', \mathrm{PC}(\mathscr{V}'), \mathscr{A}' \rangle$. Since S4' and S5' are Skolem resolutions of S4 and S5, then by Skolem's Theorem, S' is an inessential extension of S, and therefore S' is consistent if and only if S is consistent.

Let Y be the set of matrices of the members of $\mathscr{A}'$. Thus Y consists of the following five quasi-formulae:

S1''	$\neg (x < x),$
S2''	$(x < y \wedge y < z) \rightarrow x < z,$
S3''	$x < y \vee y < x \vee y = x,$

S4'' $x < gx,$
S5'' $c = y \vee c < y.$

In order to prove the consistency of S', and therefore of S, we shall describe an effective procedure for assigning truth values to all elementary formulae of $\mathscr{L}(\mathscr{V}')$ in such a way that every substitution instance of a member of Y has the value 1 ('true'). It will then follow by the First ε-Theorem that f is not a theorem of S'.

First of all, to any formula of the form $s = t$ we assign the value 1 if s and t have the same length, and the value 0 if they do not. Similarly, to any formula of the form $s < t$ we assign the value 1 if the length of s is less than the length of t, and the value 0 if it is not. Notice that one can effectively compare the lengths of any two expressions by successively crossing off the initial symbol from each of them until (at least) one of them is reduced to the empty expression. As usual, we assign the truth value 0 to the formula f. Since every atom of $\mathscr{L}(\mathscr{V}')$ is either the formula f or a formula of the form $s = t$ or $s < t$, then this assignment of values to the atoms can be extended to all elementary formulae by means of the truth functional interpretations of the propositional connectives.

We shall now show that, for any elementary formula B, if B is a theorem of S', then B has the truth value 1. By the First ε-Theorem, if B is a theorem of S', then there exists a set Z of substitution instances of the members of Y such that $Z \vdash_{EC} B$. However, it is easy to see that every member of Z must have the value 1, and furthermore every E1-axiom and every E3-axiom must have the value 1. Since the propositional axioms are tautologies and since the rule of modus ponens preserves the truth value 1, then B has the value 1. However f has the value 0. Consequently f is not a theorem of S', and therefore S' and S are consistent.

In general, we may describe the formalists' method of proving consistency as follows. Suppose $\mathscr{T}$ is some theory based on the predicate calculus. By replacing each axiom of $\mathscr{T}$ by some prenex equivalent of that axiom and then taking the Skolem resolutions of these prenex formulae and adjoining the new Skolem functions to the vocabulary, one obtains a theory $\mathscr{T}'$ which is an inessential extension of $\mathscr{T}$. One then tries to find an effective assignment of truth values to the atomic formulae of $\mathscr{T}'$ such that every E1-axiom, every E3-axiom, and every substitution instance of the matrices of the axioms of $\mathscr{T}'$ has the value 1 under this assignment. If this can be done, it then follows that both $\mathscr{T}'$ and $\mathscr{T}$ are consistent. A detailed account of this method of proof is given in Hilbert and Bernays [1939], where the principal results are embodied in the Consistency Theorem (*Widerspruchsfreiheits-Theorem*, pages 36–37).

This method of proof actually establishes more than just the consistency of $\mathscr{T}$, since one shows that every elementary theorem of $\mathscr{T}$ has an elementary proof, i.e., one without quantifiers. A major objective of Hilbert's formalist programme was to justify the use of infinity in mathematics by proving that the non-finitary statements and methods can be eliminated from proofs of finitary (i.e. elementary) statements. Kreisel [1964], page 157, explains this aspect of Hilbert's programme as follows:

'. . . The *Consistency Problem* was associated with the problem of understanding the concept of infinity. He (Hilbert) sought such an understanding in understanding the use of transfinite machinery from a finitist point of view. And this he saw in the elimination of transfinite (ε-) symbols from proofs of formulae not containing such symbols. He was convinced from the start that such an elimination was possible, and expressed it by saying that the problems of foundations were to be *removed* or that doubts were to be eliminated instead of saying that they were to be investigated.'

The classical logic of mathematics, as formalized by the predicate calculus, transcends the limits of finitary reasoning since it admits statements which refer to an infinite totality of objects. For example, if p is a prime number, the statement 'there exists a prime number which exceeds p' is non-finitary, since it asserts the existence of a number having a certain property in the infinite totality of numbers which exceed p. On the other hand, the stronger statement 'there exists a prime number between $p + 1$ and $p! + 1$' is finitary since it can be expressed as a finite disjunction (cf. Hilbert [1926]). Furthermore, certain arguments which are used in classical logic and which can be formalized in the predicate calculus are unacceptable from the finitary point of view. For example, in classical logic one can prove that there exists a number which has a certain property by deducing a contradiction from the assumption that every number does not have this property.

Perhaps, the main significance of Hilbert's First ε-Theorem is the following. Although classical logic, as formalized by the predicate calculus, contains certain non-finitary elements, any proof of a finitary statement can be converted into a finitary proof of that statement. It is in this sense that Hilbert justifies the use of the concept of infinity.

Unfortunately, this use of the First ε-Theorem has a very limited range of applications. If the formal theory under consideration is at all complicated, the First ε-Theorem may be inadequate for proving the eliminability of non-finitary elements from proofs of finitary statements. We shall consider such a theory in the following sections.

3 Formal Arithmetic

As we have already mentioned, one of the primary goals of Hilbert's formalist programme was to prove by finitary means the consistency of formal arithmetic, i.e., the theory which deals with the additive and multiplicative structure of the natural numbers. In the following sections we shall describe the way in which Hilbert hoped to use the ε-symbol to prove this consistency result.

Arithmetic can be formalized as follows. Let $\mathscr{V}$ be the vocabulary consisting of the 0-place function symbol $\bar{0}$ (which designates the natural number 0), the 1-place function symbols $'$ and g (which designate, respectively, the successor function and the predecessor function), and the 2-place function symbols $+$ and $\cdot$ (which designate addition and multiplication). As usual, we write $(t)'$, $(s + t)$, and $(s \cdot t)$ instead of $'t$, $+st$, and $\cdot st$, and we shall omit parentheses whenever their omission gives rise to no ambiguity. Let $\mathscr{A}$ be the set consisting of the following formulae:

N1 $\forall x \neg (\bar{0} = x')$
N2 $\forall x \forall y (x' = y' \to x = y)$
N3 $\forall x (x + \bar{0} = x)$
N4 $\forall x \forall y (x + (y)' = (x + y)')$
N5 $\forall x (x \cdot \bar{0} = \bar{0})$
N6 $\forall x \forall y (x \cdot (y)' = (x \cdot y) + x)$
N7 $\forall x (\bar{0} = x \vee (gx)' = x)$

and all universal closures of quasi-formulae of the form

I $A(\bar{0}) \to \forall x (A(x) \to A(x')) \to \forall x A(x).$

The form I is called the *axiom schema of induction* and any universal closure of a quasi-formula of this form is called an *induction axiom*. *Formal arithmetic*, which we denote by N, is now defined as the formal theory $\langle \mathscr{V}, \mathrm{PC}(\mathscr{V}), \mathscr{A} \rangle$. Actually, the function symbol g and axiom N7 are usually omitted in formalizations of arithmetic. We include them here only for technical reasons. Notice that formula N7 is the Skolem resolution of $\forall x \exists y (\bar{0} = x \vee y' = x)$. This latter formula can be deduced without using axiom N7.

Let $\mathfrak{N}$ be the model $\langle M, \Pi, \Phi \rangle$ for $\mathscr{V}$, where M is the set of natural numbers, Π assigns the above-mentioned interpretations to the symbols $\bar{0}$, $'$, g, $+$, and $\cdot$ and assigns the number 0 to every individual symbol, and Φ is the least number operator, i.e., for any subset M_1 of M, if M_1 is non-empty, $\Phi(M_1)$ is the least member of M_1, and if M_1 is empty, $\Phi(M_1)$ is 0. (For the interpretation of the predecessor function g we arbitrarily define the predecessor of 0 to be 0.) We shall refer to this model as the *standard model* of arithmetic. Since every axiom of N is true in $\mathfrak{N}$, then by the soundness of the predicate calculus every theorem of N is true in $\mathfrak{N}$.

Since it is possible to prove in N all the usual results about the addition and multiplication of natural numbers, the formalists felt that this theory was an adequate formalization of arithmetic and that a finitary proof of its consistency would justify the use of non-finitary statements in informal arithmetic. However, Gödel [1931] proved two profound results which under-mined the whole formalist programme. His first result, which implies that N is not an adequate formalization of arithmetic, was that if N is consistent then there exists an ε-free formula A, containing no individual symbols, such that A is true in the standard model $\mathfrak{R}$ but is not a theorem of N. Furthermore, this deficiency cannot be rectified by adding new axioms to N. His second result, a corollary of the first, showed that any proof of the consistency of N must involve techniques or concepts which cannot be formalized within N. In particular, this result ruled out the possibility of proving the consistency of N by finitary reasoning.

Despite Gödel's results it is worth while examining Hilbert's unsuccessful attempt to find a finitary consistency proof of N since his methods can be used both to demonstrate the consistency of weakened versions of N and also to prove, using a non-finitary but limited metatheory, that N itself is consistent. Furthermore, his whole approach provides us with a clear analysis of the nature of the abstract and the concrete in mathematics.

EXERCISES

1. Prove that the following formulae are theorems of N:
    (i) $\forall x \forall y (x + y = y + x)$,
    (ii) $\forall x \forall y (x \cdot y = y \cdot x)$,
    (iii) $\forall x \forall y \forall z (x \cdot (y + z) = (x \cdot y) + (x \cdot z))$.

2. For any quasi-terms s and t, let $s < t$ be an abbreviation of the quasi-formula $\exists w(s + w' = t)$, where w is some variable which does not occur free in s or t. Prove that with this definition of $<$ the axioms S1–S5 of the theory S (page 86) are theorems of N.

(For solutions of these exercises see Mendelson [1964], pages 104–112.)

3.1   Numerals and numerically true formulae

The consistency of N would be established if we could prove that every theorem of N has a certain property which the formula f does not have. If we were to allow ourselves the luxury of using non-finitary reasoning, a suitable property would be 'truth in the standard model'. Of course, from our finitary point of view this property is far too abstract. However, we shall now show that if we restrict our attention to elementary formulae and replace the abstract notion of 'natural number' by the concrete notion of 'numeral', we can define in concrete terms an effective notion of 'numerical truth'. Having defined this notion, we can then try to prove that every elementary theorem of N is numerically true, thus establishing the consistency of N.

An expression of the language $\mathscr{L}(\mathscr{V})$ is called a *numeral* if the only symbol occurring in that expression is the function symbol $'$. Thus, in particular, the empty expression e is a numeral. We shall refer to this numeral as *zero*. For any numeral n, the *successor* of n is the numeral n' and the *predecessor* of n is the numeral obtained from n by removing the last occurrence of $'$ in n. For any two numerals m and n, the *sum* of m and n is the numeral mn, and the *product* of m and n is the numeral obtained from n by replacing each occurrence of the symbol $'$ in n by the numeral m. For example, the successor of the numeral zero is $'$, the predecessor of $'''$ is $''$, the sum of $''$ and $'''$ is $'''''$, and the product of $'''$ and $''$ is $''''''$.

To each ε-free term t of the language $\mathscr{L}(\mathscr{V})$ we assign a unique numeral n, called the *numerical value* of t, as follows. To each individual symbol a and to the symbol $\bar{0}$ we assign the numeral zero. If we have already assigned the numerals m and n to the terms s and t respectively, then we assign the successor of m to the term s', the predecessor of m to the term gs, the sum of m and n to the term $s + t$, and the product of m and n to the term $s \cdot t$. Now to any ε-free formula of the form $s = t$ we assign the truth value 1 if the numerals assigned to s and t are of the same length, i.e., if s and t have the same numerical values. Otherwise, we give it the truth value 0. For example, for any ε-free terms s and t, to the formula

$$s \cdot (t)' = (s \cdot t) + s$$

is assigned the truth value 1, and to the formula

$$\bar{0}'' \cdot \bar{0}''' = \bar{0}'' + \bar{0}'''$$

is assigned the truth value 0.

Since the vocabulary $\mathscr{V}$ contains no predicate symbols, every elementary formula of $\mathscr{L}(\mathscr{V})$ is built up by means of the propositional connectives from formulae of the form $s = t$ and possibly the formula f. Consequently by using the above assignment of truth values to formulae of the form $s = t$ and by giving the symbols f, $\neg$, $\wedge$, $\vee$, and $\rightarrow$ their usual truth functional interpretations, we can assign a unique truth value to any elementary formula B. If this truth value is 1, we say that B is *numerically true*. For example, an elementary formula of the form $s' = t' \rightarrow s = t$ is numerically true, since regardless of what numerals are assigned to s and t the case where $s' = t'$ has the truth value 1 and $s = t$ has the truth value 0 can never arise. It is important to observe that this definition of numerical truth is an effective one. In other words, for any given elementary formula B, the above definitions describe an effective procedure for determining whether or not B is numerically true.

EXERCISES

1. Prove that every ε-free substitution instance of the matrices of axioms N1–N7 is numerically true.

2. Let N_1 be the theory whose axioms are the formulae N1–N7. Give a finitary proof of the consistency of this theory.

3.2   Formal arithmetic based on the ε-calculus

We now turn to the problem of proving that every elementary theorem of N is numerically true. The exercises at the end of the last section reveal that the First ε-Theorem provides a simple solution to this problem if we exclude the induction axioms from N. However, when the induction axioms are included, the problem is much more difficult, since these axioms are not ∀-prenex formulae. Although we know by Skolem's Theorem that N has an inessential extension N′ in which all the axioms are ∀-prenex formulae, in order to form N′ we must replace each induction axiom by the Skolem resolution of some prenex equivalent of that axiom and adjoin the new Skolem functions to the vocabulary $\mathscr{V}$. Obviously, the theory N′ which is obtained in this way is very complicated, and it seems unlikely that one could devise an effective notion of numerical truth for the elementary formulae of this theory.

In order to overcome this difficulty Hilbert and Bernays define a new formalization of arithmetic, which we shall denote by N_ε, which is based on the ε-calculus and in which the ε-symbol is used in effect as a least number operator. The formulation of the theory N_ε is based on the well-known fact that the principle of mathematical induction is deducible from the principle of the least number.

We can define the theory N_ε as follows. Let $\mathscr{V}$ be the vocabulary used in defining the theory N and let $\mathscr{A}_\varepsilon$ be the set of formulae of $\mathscr{L}(\mathscr{V})$ which are instances of the following axiom schemata:

N1′	$\neg(\bar{0} = t')$
N2′	$s' = t' \rightarrow s = t$
N3′	$s + \bar{0} = s$
N4′	$s + (t)' = (s + t)'$
N5′	$s \cdot \bar{0} = \bar{0}$
N6′	$s \cdot (t)' = (s \cdot t) + s$
N7′	$\bar{0} = s \vee (gs)' = s$
ε_1	$\neg A(\varepsilon x A) \rightarrow \neg A(t)$
ε_2	$t' = \varepsilon x A \rightarrow \neg A(t)$

We now define N_ε to the theory $\langle \mathscr{V}, \varepsilon(\mathscr{V}), \mathscr{A}_\varepsilon \rangle$ where as usual, $\varepsilon(\mathscr{V})$ denotes the ε-calculus for $\mathscr{V}$. A formula of the form ε_1 is called an ε_1-*axiom* and one of the form ε_2 an ε_2-*axiom*. Note that the ε_1-axioms are theorems of the ε-calculus by virtue of axiom schemata Q3 and Q4. However, it is convenient to include formulae of this form as axioms of the theory in order to prove the eliminability of the quantifiers.

Recall that the choice function Φ of the standard model $\mathfrak{N}$ was defined as the least number operator. It is easy to see that with this interpretation of the ε-symbol the ε_2-axioms are all true. Consequently all the axioms of N_ε are true in $\mathfrak{N}$, and therefore by the soundness of the ε-calculus every theorem of N_ε is also true in $\mathfrak{N}$.

However, the least number operator is not the only interpretation of the ε-symbol which would satisfy the ε_2-axioms. These axioms merely state, intuitively, that the number denoted by $\varepsilon x A$ is not the successor of a number having the property A. For example, if A is the quasi-formula

$$\exists y(y \cdot \overline{0}'' = x),$$

i.e., 'x is an even number', then there is no way of knowing which even number $\varepsilon x A$ designates, since the ε_2-axiom $\varepsilon x A = t' \rightarrow \neg A(t)$ merely states the obvious fact that the even number designated by $\varepsilon x A$ is not the successor of an even number.

Despite the fact that the ε_2-axioms do not uniquely characterize the ε-symbol as a least number operator, we shall now see that these axioms are still strong enough to provide us with the principle of mathematical induction.

THEOREM IV.1. *If B is any formula of $\mathscr{L}(\mathscr{V})$ of the form*

$$A(\overline{0}) \rightarrow \forall x(A(x) \rightarrow A(x')) \rightarrow \forall x A(x)$$

then $\mathscr{A}_\varepsilon \vdash_\varepsilon B$. Furthermore, if B is a proper formula and x does not have a free occurrence in A(x) within the scope of an ε-symbol, then $\mathscr{A}_\varepsilon \vdash_{\varepsilon*} B$.

Proof. Let s be the term $\varepsilon x \neg A(x)$ and let X be the set

$$\mathscr{A}_\varepsilon \cup \{A(\overline{0}), \forall x(A(x) \rightarrow A(x'))\}.$$

By virtue of the ∀-introduction rule and the →-introduction rule it is sufficient to prove $X \vdash_\varepsilon A(s)$. The proof is as follows, where we let t be the term gs.

(1)	$X \vdash \overline{0} = s \vee t' = s$	axiom N7'
(2)	$X \vdash A(\overline{0})$	member of X
(3)	$X \vdash (\overline{0} = s \wedge A(\overline{0})) \rightarrow A(s)$	Theorem II.22
(4)	$X \vdash \overline{0} = s \rightarrow A(s)$	tautology rule from (2) and (3)
(5)	$X \vdash \forall x(A(x) \rightarrow A(x'))$	member of X
(6)	$X \vdash A(t) \rightarrow A(t')$	∀-elimination
(7)	$X \vdash (t' = s \wedge A(t')) \rightarrow A(s)$	Theorem II.22
(8)	$X \vdash t' = s \rightarrow \neg \neg A(t)$	ε_2-axiom and definition of s
(9)	$X \vdash t' = s \rightarrow A(s)$	tautology rule from (6), (7), and (8)
(10)	$X \vdash A(s)$	tautology rule from (1), (4), and (9)

For the special case where B is a proper formula and x does not have a free occurrence in $A(x)$ within the scope of an ε-symbol, the proof is the same except that steps (3) and (7) are justified by Theorem III.5.

Often in mathematics it is simpler to use the principle of the least number than to use mathematical induction. Similarly, because of the ε_2-axioms it is often easier to prove theorems in N_ε than it is in N. For example, there exists a very simple proof in N_ε of the formula $\forall x \neg(x' = x)$. Let t be the term $\varepsilon x \neg \neg(x' = x)$. Then the formula $t' = t \to \neg \neg \neg(t' = t)$ is an ε_2-axiom. By the tautology rule $\neg(t' = t)$ is a theorem of N_ε, and by the $\forall$-introduction rule, $\forall x \neg(x' = x)$ is also a theorem. To prove that this formula is a theorem in N one must use axioms N1, N2, and an induction axiom.

Theorem IV.1 can now be used to prove that N_ε is an extension of N.

THEOREM IV.2. *If A is a theorem of* N, *then* $\mathscr{A}_\varepsilon \vdash_{\varepsilon^*} A$, *and therefore A is a theorem of* N_ε.

Proof. Let $\mathscr{D}$ be a proof of A in N. Thus $\mathscr{D}$ is a deduction in the predicate calculus of A from some finite set X where each member of X is an (ε-free) axiom of N. By Theorem III.3, there exists a deduction $\mathscr{D}_1$ of A from X in the ε^*-calculus. It is easy to see that for each B_i in X, $\mathscr{A}_\varepsilon \vdash_{\varepsilon^*} B_i$. For, if B_i is one of the axioms N1–N7 of N, then B_i follows by the $\forall$-introduction rule from an instance of the corresponding axiom schema in N_ε, and if B_i is an induction axiom, then $\mathscr{A}_\varepsilon \vdash_{\varepsilon^*} B_i$ by the $\forall$-introduction rule and Theorem IV.1. Consequently $\mathscr{A}_\varepsilon \vdash_{\varepsilon^*} A$ by Theorem II.3(ii).

EXERCISES

1. Prove that axiom schema N1′ is redundant in the theory N_ε. In other words prove that any formula of the form $\neg(\overline{0} = t')$ is deducible from the other axioms of N_ε. Using finitary reasoning prove that axiom N1 is not redundant in the theory N. (*Hint*: Use the techniques employed in proving Theorem II.15.)

2. Prove that N_ε is an inessential extension of N, or in other words, prove that for any ε-free formula A, A is a theorem of N if and only if A is a theorem of N_ε. (*Hint*: Adjoin the ι-symbol and the appropriate rules of inference to the theory N to form the theory N_ι (cf. Hilbert and Bernays [1934]). By replacing every ε-term, $\varepsilon x A$, by the ι-term 'the least x such that A' show that every ε-free theorem of N_ε is a theorem of N_ι. The desired result now follows by Hilbert and Bernays' proof of the eliminability of the ι-symbol.) Why is the Second ε-Theorem inadequate for solving this problem?

3.3   Quantifier-free proofs

We now return to the problem of proving the consistency of N. In view of Theorem IV.2, in order to prove the consistency of N it is sufficient to prove the consistency of N_ε. This result in itself does not really simplify the problem, since N_ε is also a rather complicated theory. However, we shall now show that in fact it is sufficient to prove the consistency of a quite simple sub-theory of N_ε.

A well-formed expression of $\mathcal{L}(\mathcal{V})$ is *quantifier-free* if the symbols $\forall$ and $\exists$ do not occur in that expression. A *quantifier-free proof* is now defined to be a proof in N_ε in which every formula is quantifier-free and in which the rule of relabelling bound variables may be used as a basic rule of inference. More precisely, a quantifier-free proof of B is a sequence $\langle A_1, \ldots, A_n \rangle$ of quantifier-free formulae of $\mathcal{L}(\mathcal{V})$ such that A_n is B and for each $i = 1, \ldots, n$ at least one of the following is true:

1. A_i is an instance of one of the axiom schemata P1–P10, E1, E3, N1′–N7′, ε_1, or ε_2;
2. A_i follows by modus ponens from A_j and A_k for some $j, k < i$;
3. A_i is a variant of A_j for some $j < i$.

(Recall that A is a *variant* of B if A can be obtained from B by a succession of admissible relabellings of bound variables.) Strictly speaking, a quantifier-free proof of B is not a proof *in* N_ε, since the rule of relabelling bound variables is only a derived rule of inference in the ε-calculus. However, it is obvious that any quantifier-free proof of B can be converted into a proof of B in N_ε. On the other hand, a proof in N_ε of some quantifier-free formula B cannot necessarily be converted into a quantifier-free proof of B, since within a quantifier-free proof there are no formal counterparts of the E2-axioms. Nonetheless we can prove the following weaker result which is all that is needed to simplify the problem of proving the consistency of N.

THEOREM IV.3. *For any elementary formula B, if B is a theorem of N, then there exists a quantifier-free proof of B.*

Proof. Since B is a theorem of N, then by Theorem IV.2 there exists a deduction $\mathcal{D}$ in the ε^*-calculus of B from $\mathcal{A}_\varepsilon$. We now eliminate the quantifiers from $\mathcal{D}$ as follows.

First of all we replace every occurrence of $\forall$ by $\neg \exists \neg$, thus converting every Q1-axiom and Q2-axiom into a tautology without damaging any of the other axioms, and without affecting B since B is elementary. (Recall that E2-axioms are not used in the ε^*-calculus.) In this way we obtain a proof $\mathcal{D}_1$ of B in N_ε such that no formula in $\mathcal{D}_1$ contains the symbol $\forall$.

The following procedure can now be used to eliminate every occurrence of the symbol $\exists$. We start with some quasi-formula of the form $\exists x A$ in $\mathcal{D}_1$ where A is quantifier-free. If $\varepsilon x A$ is free for x in A, we replace every occurrence of $\exists x A$ in $\mathcal{D}_1$ by $A(\varepsilon x A)$. On the other hand, suppose that $\varepsilon x A$ is not free for x in A. For example, A might be $y = \varepsilon y(y = x)$. Let $\bar{A}$ be some variant of A such that $\varepsilon x \bar{A}$ is free for x in $\bar{A}$. For example, if A is $y = \varepsilon y(y = x)$, let $\bar{A}$

be $y = \varepsilon z(z = x)$. We may now replace every occurrence of $\exists x A$ in $\mathscr{D}_1$ by $\bar{A}(\varepsilon x \bar{A})$. Successive applications of this procedure will eventually yield a sequence $\mathscr{D}_n$ of quantifier-free formulae whose last member is the original formula B. Every Q3-axiom in $\mathscr{D}_n$ will have been converted into an ε_1-axiom (or a variant of an ε_1-axiom), every Q4-axiom will have been converted into a tautology (or a variant of a tautology), and the remaining axioms will have been converted into axioms of the same form (or variants of such axioms). Consequently we obtain a quantifier-free proof of B. This completes the proof.

The following example reveals why it is necessary in the above proof to ensure that $\varepsilon x A$ is free for x in A before replacing $\exists x A$ by $A(\varepsilon x A)$. Let A be $y = \varepsilon y(y = x)$ and let A_1 be $t = \varepsilon y(y = x)$ where t is some quantifier-free term. Then the formula

$$t' = \varepsilon y \exists x A \rightarrow \neg \exists x A_1$$

is an ε_2-axiom. If we were to eliminate the symbol $\exists$ from this axiom simply by replacing $\exists x A$ by $A(\varepsilon x A)$ and $\exists x A_1$ by $A_1(\varepsilon x A_1)$ the resulting formula

$$t' = \varepsilon y A(\varepsilon x A) \rightarrow \neg A_1(\varepsilon x A_1)$$

would not be an ε_2-axiom, since a free occurrence of y in $\varepsilon x A$ becomes bound in $A(\varepsilon x A)$ and therefore the formula $\neg A_1(\varepsilon x A_1)$ is not of the form $\neg [A(\varepsilon x A)]_t^y$ as it should be.

Historical Note: Theorem IV.3 is an improvement on the corresponding result proved by Hilbert and Bernays [1939] since their ε-axioms include all formulae of the form

$$\varepsilon_3 \qquad\qquad s = t \rightarrow [\varepsilon x A]_s^y = [\varepsilon x A]_t^y$$

as well as the ε_1-axioms and ε_2-axioms. Clearly, such formulae are weaker versions of the E2-axioms.

3.4   The consistency of arithmetic

It now follows that in order to prove the consistency of arithmetic, as it is formalized by the theory N, it is sufficient to prove the following proposition.

PROPOSITION I. *For any elementary formula B of $\mathscr{L}(\mathscr{V})$, if there exists a quantifier-free proof of B, then B is numerically true.*

Once this proposition has been proved, it then follows by Theorem IV.3 that every elementary theorem of N is numerically true. Since the elementary formula f is not numerically true, then f is not a theorem of N, and N is consistent. We know by Gödel's results that the above proposition cannot be proved by finitary reasoning since we would then have a finitary proof of the consistency of arithmetic. However, Hilbert and Bernays [1939] show that a finitary proof is possible if a certain condition is imposed on the ε_2-axioms. We shall now see what that condition is and what effect it has on the original theory N.

For any ε_1-axiom $\neg A(\varepsilon x A) \rightarrow \neg A(t)$ or any ε_2-axiom $t' = \varepsilon x A \rightarrow \neg A(t)$, the term $\varepsilon x A$ is said to *belong* to that axiom. The *rank* of an ε_2-axiom is now defined to be the rank of the ε-term belonging to it. It follows from our definition of rank (page 70) that the rank of an ε_2-axiom,

$$t' = \varepsilon x A \rightarrow \neg A(t),$$

equals 1 if and only if x does not have a free occurrence in A within the scope of an ε-symbol. We can now give an exact statement of the result which Hilbert and Bernays prove.

PROPOSITION II. *For any elementary formula B of $\mathscr{L}(\mathscr{V})$, if there exists a quantifier-free proof of B in which the rank of every ε_2-axiom equals 1, then B is numerically true.*

If we carry this restriction on the ε_2-axioms back to the original theory N, it follows that N is consistent provided that we define the induction axioms to be all universal closures of quasi-formulae of the form

$$A(\overline{0}) \rightarrow \forall x(A(x) \rightarrow A(x')) \rightarrow \forall x A(x)$$

where x *does not have a free occurrence in $A(x)$ within the scope of a quantifier*. We shall refer to this weakened version of N as *restricted arithmetic*.

The proof of Proposition II is rather long and complicated, and instead of giving all the details we shall merely outline the basic ideas.

Let B be any elementary formula of $\mathscr{L}(\mathscr{V})$ and let $\mathscr{D}$ be a quantifier-free proof of B in which the rank of every ε_2-axiom equals 1. We would like to prove that B is numerically true. First of all, using essentially the same techniques which we used in proving the Rank Reduction Theorem we can eliminate from $\mathscr{D}$ all the ε-terms with rank > 1. Consequently, we may assume that the rank of every ε-term occurring in $\mathscr{D}$ equals 1. In order to prove that B is numerically true, we would like to replace each ε-term in $\mathscr{D}$ by an appropriate ε-free term and then show that this total replacement (*Gesamtersetzung*) of ε-terms by ε-free terms converts every formula in $\mathscr{D}$ into a numerically true formula. Since the instances of axiom schemata P1–P10, E1, E3, N1'–N7' are converted into numerically true formulae no matter what terms are used to replace the ε-terms, our only concern is in finding a total replacement of ε-terms which will convert the ε_1-axioms and ε_2-axioms into numerically true formulae. (Because of the rule of relabelling bound variables we must also make sure that any two ε-terms which are variants of each other are replaced by the same term. This presents no difficulty if we simply regard any two such terms as being the same term.)

To illustrate how such a total replacement can be found we shall consider the special case where no ε-term in $\mathscr{D}$ contains another ε-term as a subterm. (In the terminology of Hilbert and Bernays this is the case where the *degree* of every ε-term in $\mathscr{D}$ equals 1.) Since the rank of every ε-term in $\mathscr{D}$ equals 1, it follows from this additional restriction that if $\varepsilon x A$ occurs in $\mathscr{D}$, then A is ε-free. In order to find a total replacement of ε-terms which will convert the ε_1-axioms and ε_2-axioms into numerically true formulae we proceed by trial and error. Let R_1 be the total replacement whereby each ε-term is replaced by $\bar{0}$. Notice that every ε_2-axiom is thereby converted into a numerically true formula of the form

$$t' = \bar{0} \to \neg A(t).$$

If R_1 works, then we are done. If not, then certain of the ε_1-axioms must be to blame. Let

$$\neg A(\varepsilon x A) \to \neg A(t)$$

be one of the offending ε_1-axioms. Thus R_1 converts this axiom into the numerically false formula

$$\neg A(\bar{0}) \to \neg A(s),$$

where s is obtained from t by replacing any ε-terms in t by $\bar{0}$. Notice that the quasi-formula A is unaffected by the replacement R_1 since A is ε-free. We now compute the numerical value of s and the truth values of the formulae

$$A(\bar{0}), A(\bar{0}'), A(\bar{0}''), \ldots, A(\bar{0}^n)$$

where n is the numerical value of s. Let $A(r)$ be the first formula in the series which is numerically true. Such a formula exists since $\neg A(\bar{0}) \to \neg A(s)$ is numerically false and therefore the formula $A(\bar{0}^n)$ is numerically true. We shall refer to the term r as the *minimal value* of $\varepsilon x A$. It is easy to see that if $\varepsilon x A$ is replaced by r, then the ε_1-axioms and ε_2-axioms to which $\varepsilon x A$ belongs are converted into numerically true formulae no matter what terms are used to replace the other ε-terms. Let R_2 be the total replacement whereby $\varepsilon x A$ (and all its variants) are replaced by r and all the other ε-terms are replaced by $\bar{0}$. If R_2 works, we are done. If not, we calculate the minimal value of one of the offending ε-terms and then define a new total replacement R_3 as before. After at most $m + 1$ such attempts, where m is the number of ε-terms in $\mathscr{D}$, we arrive at a total replacement which converts every formula in $\mathscr{D}$ into a numerically true formula, thus proving that B is numerically true.

Of course the above restriction on the 'degree' of the ε-terms in $\mathscr{D}$ constitutes a very special case. In the general case the ε-free term which is used to replace one ε-term will depend on the terms which are used to replace the subterms of that term. However Hilbert and Bernays show that the basic ideas which we have just used in dealing with the special case can be applied to the general case to provide a total replacement which converts all the formulae in $\mathscr{D}$ into numerically true formulae, assuming that every ε-term in $\mathscr{D}$ has rank $= 1$. In this way they prove Proposition II and therefore the consistency of restricted arithmetic.

The basic ideas underlying Hilbert and Bernays' proof were originally devised by Ackermann [1924] in his doctoral dissertation written under Hilbert. As we have already mentioned, this was the first published work in which the ε-symbol was used. It is interesting to point out that Ackermann's dissertation was intended to prove the consistency of analysis. However, at the time of publication an error was discovered which invalidated many of the results. In order to correct the error Ackermann introduced a footnote (page 9) which severely restricted the formal theory he was dealing with. The proof of the consistency of restricted arithmetic which is given in Hilbert and Bernays [1939] and which we have just described is based on a letter from Ackermann to Bernays in which Ackermann clarifies and develops the methods used in his dissertation. Other finitary proofs of the consistency of restricted arithmetic were discovered by von Neumann [1927] and Herbrand [1931]. (For further historical details and for English translations of many important papers and lectures by Hilbert, Bernays, von Neumann, Ackermann, and Herbrand, see van Heijenoort [1967].)

The formalists' attempts to find a finitary proof of the consistency of unrestricted arithmetic came to an end in 1931 with the publication of Gödel's famous paper. By modifying Gödel's argument, Hilbert and Bernays [1939], pages 324–340, show that no proof of the consistency of N can be formalized within N. Thus any consistency proof must in some way or other involve techniques which cannot be formalized in N. Gentzen [1936] and Ackermann [1940] have constructed such consistency proofs by using transfinite induction up to the first ε-number (the first ordinal α such that $\omega^\alpha = \alpha$).

Ackermann's proof is based on an extension of his earlier methods. By using transfinite induction he essentially shows that given any array of ε_1-axioms and ε_2-axioms it is possible to assign numerals to the ε-terms so that these axioms are all converted into numerically true formulae, thus proving Proposition I and the consistency of N. An exposition of Ackermann's proof is given by Wang [1963], pages 362–370.

More recently, Tait [1965] has used recursive function theory to formalize the elimination of ε-terms from quantifier-free formalizations of arithmetic. He proves that if S is arithmetic with induction up to some ordinal ξ, then the elimination of ε-terms from proofs in S can be achieved by using second order functionals defined by transfinite recursion up to ξ, ξ^ξ, ξ^{ξ^ξ}, . . . etc. This theorem sharpens the results of Ackermann, Hilbert, and Bernays by making more explicit the metatheory which is used in eliminating the ε-symbol.

Gentzen's proof [1936] of the consistency of arithmetic depends on the eliminability of a certain rule of inference, the cut rule, and does not involve the ε-symbol. In many ways this method of proof is more straightforward than that of Ackermann. A simple version of this proof is given by Mendelson [1964], pages 258–270.

4   Theories based on the ε-calculus

The ε-calculus is often used instead of the predicate calculus in formalizing certain mathematical theories, particularly arithmetic and set theory. The advantages of using the ε-calculus in this way are many. We have already discussed how the ε-calculus can be used to prove the consistency of arithmetic. We shall now consider how the ε-symbol can simplify the actual formulation of a theory. It has been seen in Chapter II, §12 that the formulation of the underlying logic is simplified by the availability of the ε-symbol, since, for example, the ε-calculus provides simple derived rules of inference for the introduction and elimination of quantifiers. Three other advantages of using the ε-calculus as the basis of a theory are: (i) the ι-symbol is superfluous, since its role is assumed by the ε-symbol; (ii) Skolem functions can be explicitly defined as ε-terms; (iii) the ε-symbol can be used to define certain entities and concepts whose intended interpretations are to some extent indefinite.

If one is to use the ε-calculus rather than the predicate calculus in formalizing some mathematical theory one would like to know how this formalization compares with the corresponding formalization based on the predicate calculus. Obviously, the Second ε-Theorem provides the following answer to this question.

THEOREM IV.4. *Let $\mathscr{T}$ be the theory $\langle \mathscr{V}, \mathrm{PC}(\mathscr{V}), \mathscr{A} \rangle$ where every member of $\mathscr{A}$ is ε-free, and let $\mathscr{T}_\varepsilon$ be the theory $\langle \mathscr{V}, \varepsilon(\mathscr{V}), \mathscr{A} \rangle$. Then for any ε-free formula A of $\mathscr{L}(\mathscr{V})$, A is a theorem of $\mathscr{T}$ if and only if A is a theorem of $\mathscr{T}_\varepsilon$. In other words, $\mathscr{T}_\varepsilon$ is an inessential extension of $\mathscr{T}$.*

Proof. If A is a theorem of $\mathscr{T}$, then by Theorem III.2, A is a theorem of $\mathscr{T}_\varepsilon$. Conversely, if A is a theorem of $\mathscr{T}_\varepsilon$ then by the Second ε-Theorem A is a theorem of $\mathscr{T}$.

It is important to notice that the above theorem would not hold if we removed the condition that every member of $\mathscr{A}$ is ε-free, since the Second ε-Theorem could no longer be used. We shall return to this important point when we discuss formalizations of set theory which are based on the ε-calculus.

We shall now look at some of the simplifications which the ε-calculus provides.

4.1   The ι-symbol

It is often desirable to have within a formal theory $\mathscr{T}$ some way of designating 'the unique x such that A'. The ι-symbol was introduced for just this reason. The formal treatment of this symbol is given by the following ι-rule (cf. Hilbert and Bernays [1934] page 384):

If the formulae

(1) $\exists x A$
(2) $\forall x \forall y((A \wedge A_y^x) \to x = y)$

are theorems of a theory $\mathscr{T}$, then $\iota x A$ is a term of $\mathscr{T}$ and the formula

(3) $A(\iota x A)$

is a theorem of $\mathscr{T}$.

Formula (1) is called the *existence condition* and formula (2) the *uniqueness condition*.

If the theory $\mathscr{T}$ is based on the ε-calculus, then the ι-symbol and ι-rule are superfluous since one may replace $\iota x A$ by $\varepsilon x A$. In this way (3) follows from (1) by the $\exists$-elimination rule.

One objection to the above treatment of the ι-symbol is that when the ι-rule is adjoined the concept of a term becomes undecidable since there may be no way of knowing whether or not formulae (1) and (2) are theorems (cf. Bernays [1958], page 49). For this reason the following approach is often used. We write $\exists! x A$ as an abbreviation for

$$\exists x (A \wedge \forall y(A_y^x \to x = y))$$

where y is not free in A. Thus $\exists! x A$ may be read as 'there exists a unique x such that A'. Notice that $\exists! x A$ is logically equivalent to the conjunction of formulae (1) and (2). We adopt the ι-symbol as a new logical symbol of the language and we enlarge the rules of formation of the language so that for any quasi-formula A and any variable x, the expression $\iota x A$ is a well-formed quasi-term. We then adjoin all formulae of the following forms as additional axioms of the theory:

(4) $\exists! x A \to A(\iota x A)$
(5) $\neg \exists! x A \to \iota x A = t,$

where t is some specified term of the language such as the symbol $\bar{0}$ in arithmetic or the term denoting the empty set in set theory. Intuitively, these axioms say that if there exists a unique x such that A, then $\iota x A$ designates that unique object, and if not, then $\iota x A$ designates whatever t designates.

Once again, however, if our theory is based on the ε-calculus, there is no need to adjoin the ι-symbol as a new logical symbol and adopt formulae (4) and (5) as additional axioms, since the quasi ι-terms may be defined in terms of the ε-symbol as follows:

$$\iota x A =_{\text{Df}} \varepsilon x((\exists! x A \wedge A) \vee (\neg \exists! x A \wedge x = t)),$$

where t is the term in (5). In other words we regard ιxA as the formal abbreviation of the expression on the right. With this definition of the ι-symbol it is easy to prove that all formulae of the forms (4) and (5) above are theorems of the ε-calculus and hence of our theory. To see this observe that any formula of the form

$$(6) \qquad \exists x((\exists!xA \wedge A) \vee (\neg \exists!xA \wedge x = t))$$

is a theorem of the ε-calculus, and therefore by the $\exists$-elimination rule and our definition of ιxA, the formula

$$(7) \qquad (\exists!xA \wedge A(\iota xA)) \vee (\neg \exists!xA \wedge \iota xA = t)$$

is a theorem. Formulae (4) and (5) follow from (7) by the tautology rule.

EXERCISE

Prove that any formula of the form (6) is a theorem of the ε-calculus.

4.2   Skolem functions

In §2 of this chapter we saw that by replacing the axioms of a theory $\mathscr{T}$ by their Skolem resolutions and by adjoining the new Skolem functions to the vocabulary one can sometimes find a proof of the consistency of $\mathscr{T}$. This elimination of the existential quantifiers from the axioms of $\mathscr{T}$ also provides a more practical formulation of the theory itself. For example, in set theory instead of stating the power set axiom in the usual existential form,

$$(1) \qquad \forall x \exists y \forall z (z \in y \leftrightarrow z \subseteq x),$$

it is preferable to state it in the form

$$(2) \qquad \forall x \forall z (z \in \pi(x) \leftrightarrow z \subseteq x),$$

where the new function symbol π is taken as a primitive (cf. the system of Bernays [1958]). The availability of this symbol makes it possible to designate within the theory the power set of any given set. If, however, the theory is based on the ε-calculus, there is no need to take the formula (2) instead of (1) as an axiom, since we can define $\pi(t)$ for any t as follows

$$\pi(t) =_{\mathrm{Df}} \varepsilon y \forall z (z \in y \leftrightarrow z \subseteq t).$$

Under this definition of π, formula (2) follows from (1) by the $\exists$-elimination rule. Thus, in general, if a theory $\mathscr{T}_\varepsilon$ is based on the ε-calculus, the axioms may be stated in the weaker existential form (or unresolved form) and the Skolem functions for the existential variables may be introduced by explicit definitions in terms of the ε-symbol. By Theorem IV.4 it follows that the introduction of these new function symbols does not essentially strengthen the theory, since any formula not containing these symbols which is a theorem of $\mathcal{T}_\varepsilon$, is a theorem of the corresponding theory $\mathcal{T}$ which is based on the predicate calculus (assuming that the axioms of $\mathcal{T}_\varepsilon$ are ε-free).

This method of eliminating existential quantifiers by means of ε-terms can be used to simplify not only the axioms of $\mathcal{T}_\varepsilon$, but also the theorems. For example, suppose the formula

(3) $$\exists y \forall z(z \in y \leftrightarrow B)$$

is a theorem of some formalization of set theory, where B is a quasi-formula not containing a free occurrence of y. Formula (3) asserts that there exists a set y whose members are those entities which satisfy B. Using the terminology of Bourbaki (1954), we say 'B is collectivizing in z'. It is convenient, though not necessary, to have some formal apparatus for designating this set. Such designations are possible if, for example, the primitive symbols of the theory include the 'comprehension operator' $\hat{\,}$, where the term $\hat{z}B$ denotes the set whose members are the entities which satisfy B. If, however, the ε-symbol is available, the operator $\hat{\,}$ need not be taken as a primitive, but can be defined explicitly as follows

(4) $$\hat{z}B =_{\mathrm{Df}} \varepsilon y \forall z(z \in y \leftrightarrow B),$$

where B is any quasi-formula and y any variable which does not occur free in B. Then by formula (3) and the $\exists$-elimination rule we obtain the formula

(5) $$\forall z(z \in \hat{z}B \leftrightarrow B),$$

which asserts that the term $\hat{z}B$ does designate the required set. It is interesting to note that definition (4) may be used for any quasi-formula B, even if B is not collectivizing in z. For example, B may be the quasi-formula $\neg z \in z$. This definition of $\hat{z}B$ does not introduce any contradictions, such as Russell's paradox, since formula (5) depends on (3), i.e. on the fact that B is collectivizing in z. If B is not collectivizing in z, the expression $\hat{z}B$ as defined by (4) is still a well-formed term of the language, but nothing very much can be said about it (cf. Bourbaki [1954] p. 63). In this case $\hat{z}B$ is a 'null term' (see Chapter I, page 54).

We shall return to this subject in a later section (see page 107) where a formal system similar to that used by Bernays [1958] is presented in which the operator $\hat{\,}$ is taken as a primitive symbol and Church's schema

$$\forall z(z \in \hat{z}B \leftrightarrow B)$$

is taken as an axiom schema.

4.3 Definability of indeterminate concepts

The above definition of the comprehension operator is used by both Bourbaki [1954], p. 63, and Ackermann [1937–8]. Clearly, the ι-symbol could be used instead of the ε-symbol in formulating this definition, since by the axiom of extensionality,

$$\forall x \forall y(x = y \leftrightarrow \forall z(z \in x \leftrightarrow z \in y)),$$

formula (3) satisfies the uniqueness condition. However, if an existential formula $\exists x A$ does not satisfy the uniqueness condition, then in order to designate some object which satisfies A, one must use the ε-symbol. In this case the designation, $\varepsilon x A$, has a certain degree of indeterminacy since as we have observed before, nothing definite can be said about $\varepsilon x A$ except that it satisfies A if anything does. It more than one entity satisfies A, then there is no way of knowing exactly which of these objects $\varepsilon x A$ designates. Occasionally, it is desirable to define objects which have just this degree of indeterminacy. We shall now consider two such occasions.

In set theory, the concept of cardinal number is difficult to formulate explicitly since the intended interpretation of this concept is rather indefinite. All that we require of the definition is that the following formula be a theorem

(1) $$\forall x \forall y(x \sim y \leftrightarrow \bar{x} = \bar{y}),$$

where $\bar{x}$ and $\bar{y}$ are, respectively, the cardinal numbers of x and y, and the expression $x \sim y$ is an abbreviation of the assertion that there exists a one-to-one correspondence between x and y. Various definitions of $^{=}$ can be found in the literature, but probably the simplest is the following, which is used by Bourbaki and Ackermann:

(2) $$\bar{t} =_{\text{Df}} \varepsilon z(z \sim t),$$

where t is any quasi term and z does not occur free in t. Using the fact that $\sim$ is an equivalence relation, it is a simple matter to prove (1). The proof is as follows. Let s and t be any two terms. Since $t \sim t$, then $\exists z(z \sim t)$, and consequently, $\varepsilon z(z \sim t) \sim t$, i.e.

(3) $$\bar{t} \sim t.$$

Similarly, we get

(4) $$\bar{s} \sim s.$$

From (3) and (4) and the fact that $\sim$ is an equivalence relation we obtain

(5) $$\bar{s} = \bar{t} \rightarrow s \sim t.$$

On the other hand, the fact that $\sim$ is an equivalence relation implies

(6) $$s \sim t \rightarrow \forall z(z \sim s \leftrightarrow z \sim t).$$

By axiom schema E2, we obtain

(7) $$\forall z(z \sim s \leftrightarrow z \sim t) \rightarrow \varepsilon z(z \sim s) = \varepsilon z(z \sim t).$$

Therefore, (6) and (7) yield

(8) $$s \sim t \rightarrow \bar{s} = \bar{t}.$$

Consequently, from (5) and (8) we get

$$(9) \qquad\qquad s \sim t \leftrightarrow \bar{s} = \bar{t}.$$

Formula (1) now follows from (9) by the $\forall$-introduction rule.

This definition of cardinal number is an essentially indeterminate one since nothing can be said about the set $\bar{t}$ in (2) except that it is equivalent to t and that it equals the cardinal number of any set which is equivalent to t. It is just this degree of indeterminacy which we want the concept of cardinal number to possess. Ackermann [1937–8] has observed that for any equivalence relation $\sim$, definition (2) can be used to specify a representative element from each equivalence class of $\sim$. This method of designating a completely arbitrary representative of an equivalence class could have useful applications in the formulation of various mathematical theories, particularly in the introduction of 'definitions by abstraction' (cf. Beth [1959], pp. 91–95).

The basic indeterminacy of ε-terms is also used by Carnap [1961] to overcome certain difficulties which arise in the formulation of theories of empirical science. Such theories include certain terms, called the 'theoretical terms' or 'T-terms', which represent the 'unobservables' of the theory (e.g. 'temperature', 'electric field', etc.). The interpretation of these terms is provided by the postulates of the theory. These postulates are of two kinds, the 'theoretical postulates' ('T-postulates') and the 'correspondence postulates' ('C-postulates'). However, these postulates do not provide a complete interpretation of the T-terms 'because the scientist can always add further C-postulates (e.g., operational rules for T-terms) or T-postulates and thereby increase the specification of the meanings of the T-terms'. Because of the indeterminacy of these terms, the following problem arises: how can one give explicit definitions of the T-terms which satisfy the postulates without contributing anything new to the factual content of the theory? Carnap solves this problem by defining the T-terms as ε-terms, thereby obtaining just the intended degree of indeterminacy.

4.4   The ε-symbol and the axiom of choice

Finally, the indeterminate nature of the ε-symbol helps to explain the close connection which exists between this symbol and the axiom of choice. The axiom of choice differs from the other axioms of set theory in that these other axioms (e.g., the axioms of power set, pairing, replacement, etc.) not only assert the existence of a new set, but also specify the members of this set. The axiom of choice, on the other hand, merely asserts the existence of a selection set y for a given set x without actually specifying the members of y. Similarly, the quasi ε-term $\varepsilon u(u \in w)$ expresses a choice function for the set x as the variable w ranges over x, but there is no way of knowing which member of w is being selected. For these reasons the ε-symbol and Q4 axioms are often regarded as logical counterparts of the axiom of choice. However, it does not necessarily follow that the axiom of choice is derivable in a formalization of set theory which is based on the ε-calculus. For, suppose that $\mathcal{T}$ is a set theory based on the predicate calculus, and $\mathcal{T}_\varepsilon$ is the corresponding theory based on the ε-calculus, i.e. $\mathcal{T}$ and $\mathcal{T}_\varepsilon$ have the same set of axioms. Then by Theorem IV.4 (or the Second ε-Theorem), if the axiom of choice is not a theorem of $\mathcal{T}$, then it is not a theorem of $\mathcal{T}_\varepsilon$, even though this latter theory is based on the ε-calculus.

The intuitive explanation of why the ε-symbol and ε-axioms do not necessarily yield the axiom of choice is as follows. Although the quasi ε-term $\varepsilon u(u \in w)$ can be used to make a simultaneous selection from each member of a given set x, it does not necessarily follow that there exists a set y consisting of these selected entities. The axiom of choice, on the other hand, does assert the existence of the selection set y. Wang [1955] has observed that if the formula

(1) $$\forall x \exists y \forall z (z \in y \leftrightarrow \exists w(w \in x \land z = \varepsilon u(u \in w)))$$

is a theorem of $\mathcal{T}_\varepsilon$, then the ε-axioms do yield the axiom of choice. Clearly, formula (1) asserts that for any x, there exists a set y whose members are the selected entities from each member of x.

In most set theories, the axioms include the instances of a certain axiom schema, usually referred to as the axiom schema of replacement. If the theory is based on the ε-calculus, then the question of the deducibility of the axiom of choice usually hinges on whether the axioms include *all* the instances of this schema, or just those instances which are ε-free. In the former case, the ε-symbol does provide the axiom of choice, but in the latter case it does not.

For example, consider the set theory of Bourbaki [1954]. This system is based on an ε-calculus which is virtually equivalent to ours. (The differences are that the Greek letter τ, instead of ε, is used for the selection operator, and the quantifiers are defined in terms of τ instead of being taken as primitive symbols.) The theory has the following axiom schema:

S8 $\qquad \forall w \exists y \forall z (A \to z \in y) \to \forall x \exists y \forall z (z \in y \leftrightarrow \exists w(w \in x \land A))$,

where A is any quasi-formula not containing free occurrences of x and y, and w, x, y, and z are all distinct. Letting A be the quasi-formula $z = \varepsilon u(u \in w)$, S8 yields the following theorem (cf. Bourbaki, p. 66, C53):

$$\forall x \exists y \forall z (z \in y \leftrightarrow \exists w(w \in x \land z = \varepsilon u(u \in w))).$$

Consequently, the axiom of choice is derivable in this system. On the other hand, if the axiom schema S8 were subject to the restriction that the quasi-formula A must be ε-free, then all the axioms would be ε-free, and by the Second ε-Theorem the axiom of choice would not be derivable in this system (since it is known that this axiom is independent of the other axioms of a set theory based on the predicate calculus).

Wang [1955], pp. 66–67 points out another distinction between the ε-symbol and the axiom of choice:

'There are also cases where, although the ε-rule would yield the desired result, the axiom of choice would not. For example, in the Zermelo theory we can infer "$(x)R(x,εyRxy)$" from "$(x)(\exists y)Rxy$" by the ε-rule, but we cannot infer "there exists f, $(x)R(x,fx)$" from "$(x)(\exists y)Rxy$" by the axiom of choice, on account of the absence of a universal set in Zermelo's theory.'

In other words, the ε-symbol provides us with a 'universal choice function' which is defined on the class of all sets. The existence of such a function is normally provided by only the strongest forms of the axiom of choice (e.g., axiom E in Gödel [1940] and axiom A_σ in Bernays [1958]). We shall return to this subject in § 5.2.

In conclusion, if $\mathcal{T}_\varepsilon$ is some theory based on the ε-calculus, then various simplifying definitions and processes can be formulated within $\mathcal{T}_\varepsilon$, which could not be formulated within the corresponding theory $\mathcal{T}$, based on the predicate calculus. However, if all the axioms of $\mathcal{T}_\varepsilon$ are ε-free, then by the eliminability of the ε-symbol, every ε-free theorem of $\mathcal{T}_\varepsilon$ is also a theorem of $\mathcal{T}$. Thus the introduction of the ε-symbol and the ε-axioms can simplify the formulation of a theory without enlarging its set of theorems in any essential way, and in particular, without introducing any inconsistency.

On the other hand, if the axioms of $\mathcal{T}_\varepsilon$ include *all* the instances of some axiom schema, such as the axiom schema of replacement in set theory, then the Second ε-Theorem is not applicable and the theory $\mathcal{T}_\varepsilon$ is not necessarily consistent relative to the consistency of $\mathcal{T}$. Nevertheless, we shall show in § 5.2 that in the case of set theory the relative consistency of $\mathcal{T}_\varepsilon$ with respect to $\mathcal{T}$ still holds. This provides a positive solution to the following problem, raised by Fraenkel and Bar-Hillel [1958], p. 185:

'This relative consistency need no longer hold if the axioms of the theory do also contain ε-terms; indeed, the consistency of every set theory in which the axiom schema of comprehension (in any of its variants) is to hold also for conditions containing ε-terms relative to that in which this is not assumed, has not yet been proved.'

5.1   The predicate calculus with class operator and choice function

In § 4.2 we observed that in a set theory based on the ε-calculus the operator $\hat{\ }$ can be defined in terms of the ε-symbol in such a way that the formula

(1) $\forall x(x \in \hat{x}A \leftrightarrow A)$

is derivable from

$$\exists y \forall x (x \in y \leftrightarrow A).$$

It is often convenient, however, to take the symbol $\hat{}$ as a primitive and to adopt (1) as an axiom schema. In such a system the expressions of the form $\hat{x}A$, which we call class-terms, must be treated with some caution so as to avoid the paradoxes. One common way of doing this is to formulate the grammar of the language in such a way that class-terms may only appear on the right hand side of the membership symbol $\in$ (cf. Bernays [1958] p. 48). Such a system can be further strengthened by taking as a primitive a choice function whose arguments are class terms. For example, Bernays [1958] takes the symbol σ as a primitive and adopts the following formulae as axioms:

A_σ'	$a \in C \rightarrow \sigma(C) \in C,$
A_σ''	$A \equiv B \rightarrow \sigma(A) = \sigma(B),$

where a is a free set variable, and A, B, and C are free class variables. In Bernays' system of set theory, these two axioms provide a very strong form of the axiom of choice.

In this section we define a formal system, $C(\mathscr{V})$, which incorporates the class operator $\hat{}$ and the choice function σ, and investigate the relationship between this system and the ε-calculus.

We first define the *class-language*, $\mathscr{L}_1(\mathscr{V})$, which is determined by a given vocabulary $\mathscr{V}$. Let $\mathscr{V}$ be any vocabulary. The set of symbols of $\mathscr{L}_1(\mathscr{V})$ is the same as that of $\mathscr{L}(\mathscr{V})$ except that we exclude the ε-symbol and include the symbols $\in$, $\hat{}$, and σ as logical symbols. The rules of formation for defining the quasi-terms and quasi-formula of the language are the same as for $\mathscr{L}(\mathscr{V})$ (see page 11) except that we replace G8 (the rule for forming quasi ε-terms) by the following:

G8°. If t is a quasi-term, A a quasi-formula, and x any variable, then $t \in \hat{x}A$ is a quasi-formula, and $\sigma(\hat{x}A)$ is a quasi-term.

Any expression of the form $\hat{x}A$ is called a *quasi class-term*. The quasi class-terms are not included among the quasi-terms of the language, and to avoid any confusion we shall refer to the quasi-terms of the language as *quasi set-terms*. As can be seen from the above rule a quasi class-term $\hat{x}A$ can occur within a well-formed expression in only two possible contexts: (i) $t \in \hat{x}A$ and (ii) $\sigma(\hat{x}A)$. As usual, we shall use the letters s and t to denote arbitrary quasi set-terms. The quasi class-terms will be denoted by the letters S and T. For any two quasi class-terms S and T we write

$$(S = T) \quad \text{for} \quad \forall z(z \in S \leftrightarrow z \in T),$$

where z is some variable which does not occur free in S or T.

Free and bound occurrences of variables are defined in the usual way, where any occurrence of x in $\hat{x}A$ is bound, and we define a *set-term, class-term,* and *formula* as a quasi set-term, quasi class-term, or quasi-formula, respectively, in which no variable occurs free.

We now define the formal system $C(\mathscr{V})$, called the *class calculus* for the vocabulary $\mathscr{V}$. The *(logical) axioms* of $C(\mathscr{V})$ are all formulae of $\mathscr{L}_1(\mathscr{V})$ which are instances of axiom schemata P1–P10, Q1–Q3, E1, E3 (see page 39) as well as the following three schemata:

CS $\qquad\qquad\qquad t \in \hat{x}A \leftrightarrow A_t^x,$
$\sigma 1$ $\qquad\qquad\qquad \exists x(x \in S) \to \sigma(S) \in S,$
$\sigma 2$ $\qquad\qquad\qquad S = T \to \sigma(S) = \sigma(T).$

The rules of inference of the class calculus are the same as for the predicate calculus, namely, the modus ponens rule and the $\exists$-rule. The deductions in $C(\mathscr{V})$ are defined exactly as they were for the predicate calculus by first defining a derivation (see page 60). To denote that A is deducible from X in $C(\mathscr{V})$ we write $X \vdash_{C(\mathscr{V})} A$, or just $X \vdash_C A$. The usual derived rules of inference can be established for $C(\mathscr{V})$ as they were for $PC(\mathscr{V})$.

THEOREM IV.5. *Any formulae of the following forms are theorems of the class calculus*:

(i) $\qquad\qquad\qquad \exists xA \to A(\sigma(\hat{x}A)),$
(ii) $\qquad\qquad\qquad \forall z(A_z^x \leftrightarrow B_z^y) \to \sigma(\hat{x}A) = \sigma(\hat{y}B).$

Proof. Throughout the proof we shall write $\vdash$ instead of $\vdash_C$.
(i): Let a be some individual symbol not appearing in A.

$\qquad\qquad \vdash a \in \hat{x}A \leftrightarrow A_a^x$ $\qquad\qquad$ CS
$\qquad\qquad \vdash a \in \hat{x}A \to \exists x(x \in \hat{x}A)$ $\qquad\qquad$ Q3 and P3
$\qquad\qquad \vdash \exists x(x \in \hat{x}A) \to \sigma(\hat{x}A) \in \hat{x}A$ $\qquad\qquad$ $\sigma 1$
$\qquad\qquad \vdash \sigma(\hat{x}A) \in \hat{x}A \leftrightarrow A(\sigma(\hat{x}A))$ $\qquad\qquad$ CS
$\qquad\qquad \vdash A_a^x \to A(\sigma(\hat{x}A))$ $\qquad\qquad$ tautology rule
$\qquad\qquad \vdash \exists xA \to A(\sigma(\hat{x}A))$ $\qquad\qquad$ $\exists$-rule.

(ii): Let a be some individual symbol not appearing in A or B. By axiom schemata Q1 and Q3,

$$\vdash \forall z(A_z^x \leftrightarrow B_z^y) \to (A_a^x \leftrightarrow B_a^y).$$

But by CS,

$$\vdash a \in \hat{x}A \leftrightarrow A_a^x,$$
$$\vdash a \in \hat{y}B \leftrightarrow B_a^y.$$

Hence, by the tautology rule,

$$\vdash \forall z(A_z^x \leftrightarrow B_z^y) \to (a \in \hat{x}A \leftrightarrow a \in \hat{y}B).$$

By the $\forall$-rule and the definition of $=$,

$$\vdash \forall z(A_z^x \leftrightarrow B_z^y) \rightarrow \hat{x}A = \hat{y}B.$$

Consequently, by $\sigma 2$,

$$\vdash \ z(A_z^x \leftrightarrow B_z^y) \rightarrow \sigma(\hat{x}A) = \sigma(\hat{y}B).$$

The above theorem indicates that there is a close connection between the system $C(\mathscr{V})$ and the ε-calculus. Of course, this relationship is not unexpected in view of our semantic interpretation of the ε-symbol as a choice function. In order to investigate this relationship more rigorously we define a transform operation which translates a formula of $\mathscr{L}(\mathscr{V})$ into a formula $\mathscr{L}_1(\mathscr{V})$ and another transform operation which translates a formula $\mathscr{L}_1(\mathscr{V})$ into a formula of $\mathscr{L}(\mathscr{V})$.

For any formula B of $\mathscr{L}(\mathscr{V})$ the C-*transform* of B is that formula of $\mathscr{L}_1(\mathscr{V})$ which is obtained from B by replacing each quasi ε-term εxA in B by $\sigma(\hat{x}A)$. Conversely, for any formula B of $\mathscr{L}_1(\mathscr{V})$ the ε-*transform* of B is that formula of $\mathscr{L}(\mathscr{V})$ which is obtained from B by eliminating the quasi class-terms in B as follows: (i) replace each expression of the form $(t \in \hat{x}A)$ by A_t^x (if t is not free for x in A, relabel the bound variables in A so that it is free for x); (ii) replace each expression of the form $\sigma(\hat{x}A)$ by εxA.

THEOREM IV.6. *The ε-transform of an instance of* CS, $\sigma 1$, *or* $\sigma 2$ *is a theorem of the ε-calculus.*

Proof. (i): Clearly, the ε-transform of an instance of CS is a theorem of the ε-calculus since it has the form

$$\forall y(A \leftrightarrow A).$$

(ii): An instance of $\sigma 1$ has the form

(1) $$\exists x(x \in \hat{y}B) \rightarrow \sigma(\hat{y}B) \in \hat{y}B.$$

The ε-transform of (1) has the form

(2) $$\exists x[B']_x^y \rightarrow [B']_{\varepsilon y B'}^y,$$

which by the Q4 axiom schema is clearly a theorem of the ε-calculus.
(iii): An instance of $\sigma 2$ has the form

$$\forall z(z \in \hat{x}A \leftrightarrow z \in \hat{y}B) \rightarrow \sigma(\hat{x}A) = \sigma(\hat{y}B).$$

Clearly, by axiom schema E2, the ε-transform of this is a theorem of the ε-calculus.

THEOREM IV.7. (i) *If B is a theorem of the class calculus, then the ε-transform of B is a theorem of the ε-calculus.*

(ii) *If B is a theorem of the ε-calculus, then the C-transform of B is a theorem of the class calculus.*

Proof. (i): The proof follows immediately by induction on the length of the proof of B in $C(\mathscr{V})$ since by Theorem IV.6 the ε-transform of every axiom of $C(\mathscr{V})$ is a theorem of the ε-calculus, modus ponens is a rule of the ε-calculus, and the ∃-rule is a derived rule in the ε-calculus.

(ii): Similarly, the proof follows immediately by induction on the length of the proof of B, since the C-transform of every axiom of the ε-calculus is a theorem of $C(\mathscr{V})$ by Theorem IV.5, and modus ponens is a rule of inference of $C(\mathscr{V})$.

THEOREM IV.8 (Eliminability of ⌃ and σ). *Suppose X is a set of formulae of $\mathscr{L}_1$ and A is a formula of $\mathscr{L}_1$, such that the symbols ⌃ and σ do not appear in A or in any member of X. If $X \vdash_C A$, then $X \vdash_{PC} A$.*

Proof. Since $X \vdash_C A$, then there exist formulae $B_1, \ldots, B_n$ of X such that $B_1 \rightarrow \ldots \rightarrow B_n \rightarrow A$ is a theorem of $C(\mathscr{V})$. We shall denote this formula by B. Since B does not contain ⌃ or σ, then the ε-transform of B is B, and therefore by Theorem IV.7 B is a theorem of the ε-calculus. Finally by the Second ε-Theorem B is a theorem of the predicate calculus. Hence $X \vdash_{PC} A$.

The class calculus can be used as the logical basis for formalizing set theory. In this case the rules of formation of $\mathscr{L}$ are extended to include quasi formulae of the form $s \in t$. (Alternatively, two distinct symbols $\in$ and η can be used for set membership and class membership, respectively.) Although the symbol σ can have only quasi class-terms as its arguments, we can extend the use of this symbol in practice by writing

$$\sigma(t) \quad \text{for} \quad \sigma(\hat{x}(x \in t)),$$

for any quasi set-term t, where x does not occur free in t. Using Theorem IV.5, it follows that for any set-terms s and t:

$$\vdash \exists x(x \in t) \rightarrow c(t) \in t,$$
$$\vdash s = t \rightarrow \sigma(s) = \sigma(t).$$

Although the symbol σ provides the system with a 'universal choice function', in view of Theorem IV.8 it does not necessarily follow that the axiom of choice is a theorem of set theory based on the class calculus. Clearly, the relationship between the symbol σ and the axiom of choice is similar to the relationship between the ε-symbol and the axiom of choice (cf. §4.4). If the axioms of the theory include those instances of the axiom schema of replacement which contain σ, then the axiom of choice is derivable. For, the formula

(1) $$\forall x \exists y \forall z(z \in y \leftrightarrow \exists w(w \in x \wedge z = \sigma(w)))$$

together with $\sigma 1$ and $\sigma 2$ does yield the axiom of choice, and (1) is an immediate consequence of the unlimited axiom schema of replacement. In the set theory of Bernays [1958], the axioms A_σ' and A_σ'' do yield the axiom of choice because the instances of his axiom schema A3 include formulae which contain the symbol σ.

5.2   The relative consistency of set theory based on the ε-calculus

We now turn to the problem raised at the end of § 4.4—namely, whether a set theory based on the ε-calculus with an unrestricted axiom schema of replacement is consistent relative to the consistency of the same theory based on the predicate calculus.

Let ZF be a set theory, based on the predicate calculus using set-variables only (i.e., no class-terms or class-variables), whose axioms consist of the axioms of extensionality, pairing, union-set, power-set, infinity, foundation, and the axiom schema of replacement. Let ZF_ε be the corresponding theory based on the ε-calculus, where an instance of the axiom schema of replacement may contain the ε-symbol. We wish to prove that if ZF is consistent, then ZF_ε is also consistent.

Let B_σ' be the system of Bernays [1958] with axioms A_σ' and A_σ''. Let B_σ be the system of Bernays where the axioms A_σ' and A_σ'' are replaced by

$$A_\sigma \qquad\qquad a \neq 0 \rightarrow \sigma(a) \in a.$$

(In B_σ the arguments of the primitive symbol σ are set-terms, and in B_σ' the arguments are class-terms.) Bernays has shown (pp. 200–207) that under a suitable definition of $\sigma(A)$, axioms A_σ' and A_σ'' are derivable from A_σ and the axiom of foundation.

Let G′ be the set theory of Gödel [1940] with axioms A, B, C, D, and E, and let G be the same theory without Axiom E. Axiom E, the axiom of choice, is the formula

$$\exists A(\mathfrak{Un}(A) \wedge \forall x(\neg \mathfrak{Em}(x) \rightarrow \exists y(y \in x \wedge \langle yx \rangle \in A))),$$

where

$$\mathfrak{Un}(A) =_{Df} \forall u \forall v \forall w((\langle vu \rangle \in A \wedge \langle wu \rangle \in A) \rightarrow v = w),$$
$$\mathfrak{Em}(x) =_{Df} \forall u \neg (u \in x),$$

and A is a class-variable.

Suppose that B is an ε-free theorem of ZF_ε. It can be shown that the C-transform of any axiom of ZF_ε is a theorem of B_σ'. Hence by Theorem IV.7(ii), B is a theorem of B_σ'. Consequently, using Bernays' definition of σ, B is a theorem of B_σ. It can then be shown that B is a theorem of G′. Consequently, if G′ is consistent, then ZF_ε is also consistent. But by Gödel's proof [1940] of the relative consistency of the axiom of choice, G' is consistent if G is. Finally, using the method of Shoenfield [1954], G is consistent relative to the consistency of ZF. This completes the argument.

One further question remains. Is every ε-free theorem of ZF_ε also a theorem of ZF with the axiom of choice? Or, in other words, are the ε-axioms and the unrestricted axiom schema of replacement equivalent to the axiom of choice and the axiom schema of replacement? The answer to this problem seems to depend on whether the universal or the local version of the axiom of choice is used. These two versions of the axiom are defined by Lévy [1961] as follows. The local version of the axiom of choice is the formula

(1) $$\forall x \exists f \forall y (y \in x \to y = 0 \vee f'y \in y).$$

The universal axiom of choice is obtained by taking the formula

(2) $$\forall x (x \neq 0 \to \sigma(x) \in x)$$

as an axiom and allowing the instances of the axiom schema of replacement to include the primitive symbol σ. Let ZF' be the theory obtained from ZF by adjoining the local version of the axiom of choice, and let ZF_σ be the theory obtained from ZF by adjoining the universal version. The above results can be used to show that the set of ε-free theorems of ZF_ε coincides with the set of σ-free theorems of ZF_σ. However, it is not known whether this set coincides with the set of theorems of ZF'. For a partial solution of this problem, see Lévy [1961].

CHAPTER V

THE CUT ELIMINATION THEOREM

1 The sequent calculus

In this chapter we define a new formal system called the sequent calculus. This system resembles Gentzen's system LK [1934–5] except that it incorporates the ε-symbol and the identity symbol. The main theorem of this chapter, the Cut Elimination Theorem can be regarded as an analogue of Gentzen's *Hauptsatz*. We shall use this theorem to provide new proofs of the ε-Theorems (except that in the case of the Second ε-Theorem we consider an ε-calculus in which the E2-axioms are not used). Although we are in effect only giving alternative proofs of some of the theorems of Chapter III, this new approach sheds some light on the relationship between Gentzen's *Hauptsatz* and Hilbert's ε-Theorems.

Throughout this chapter $\mathscr{V}$ denotes some arbitrary vocabulary, and $\mathscr{L}$ denotes the language determined by $\mathscr{V}$.

In the sequent calculus, the rules of inference apply not to formulae, but rather to more complicated formal objects called 'sequents'. A *sequent* of the language $\mathscr{L}$ is an expression of the form $A_1, \ldots, A_n$ where the A_i are formulae of $\mathscr{L}$, $n \geqslant 0$, and the symbol ',' (comma) is a formal separation symbol of the language $\mathscr{L}$ (page 10). A formula *belongs to*, or *is a member of*, a sequent $A_1, \ldots, A_n$ if it is one of the A_i. A sequent is ε-free if every formula belonging to that sequent is ε-free. The capital Greek letters Γ, Δ, Θ, and Λ are used as syntactic variables for sequents. If Γ and Δ denote two non-empty sequents, then clearly Γ, Δ denotes a sequent. By an abuse of notation we shall also write Γ, Δ when either Γ or Δ is empty. If Γ is empty, then Γ, Δ denotes the sequent Δ, and vice versa when Δ is empty. Notice that a formula is a sequent. This situation should not lead to any confusion, however. For any sequent Δ, we write Δ^* to denote the set of formulae which belong to Δ. Thus if Δ is A, B, A, A, then $\Delta^* = \{A, B\}$.

A non-empty sequent $A_1, \ldots, A_n$ has the same semantic interpretation as the formula $A_1 \wedge \ldots \wedge A_n$. The empty sequent is given the truth value 1 (true). We shall not, however, be concerned with the semantics of sequents except as an intuitive guide to our understanding of the axioms and rules of inference of the system.

The sequent calculus deals with refutations rather than deductions. Intuitively, we can regard a refutation of Δ as a formal demonstration of the inconsistency of the set Δ^*. Each axiom is an invalid sequent, and each rule of inference leads from invalid sequents to invalid sequents. When we have defined the notion of a refutation, we can then define a 'deduction' of A from X as a refutation of some sequent $\neg A, \Delta$ where $\Delta^* \subseteq X$. In particular a 'proof' of A is a refutation of the sequent $\neg A$.

2 The axioms and rules of inference of the sequent calculus

A sequent Δ is an *axiom* of the sequent calculus if either

(i) the formula f belongs to Δ, or
(ii) for some atom A, both $\neg A$ and A belong to Δ, or
(iii) a formula of the form $\neg(t = t)$ belongs to Δ.

If Δ satisfies either (i) or (ii) above, it is called a *C-axiom*.

In defining the rules of inference we use the customary schematic notation. Thus to denote that Γ 'follows' by some rule of inference from Δ we write

$$\frac{\Delta}{\Gamma}$$

and to denote that Γ 'follows' by some rule of inference from Δ_1 and Δ_2 we write

$$\frac{\Delta_1 \quad \Delta_2}{\Gamma}$$

The sequent written below the line is called the *conclusion* of that rule of inference, and the formula(e) written above the line the *premiss(es)*.

The *rules of inference* are as follows:

$$\neg\neg\text{-rule} \quad \frac{A, \Gamma}{\neg\neg A, \Gamma}$$

$$\alpha_1\text{-rule} \quad \frac{\alpha_1, \Gamma}{\alpha, \Gamma}$$

$$\alpha_2\text{-rule} \quad \frac{\alpha_2, \Gamma}{\alpha, \Gamma}$$

$$\beta\text{-rule} \quad \frac{\beta_1, \Gamma \quad \beta_2, \Gamma}{\beta, \Gamma}$$

$$\gamma\text{-rule} \quad \frac{\gamma(t), \Gamma}{\gamma, \Gamma}$$

$$\delta\text{-rule} \quad \frac{\delta(\varepsilon\delta), \Gamma}{\delta, \Gamma}$$

$$\text{E1-rule} \quad \frac{\neg(s = t), \Gamma \quad \neg A_s^x, \Gamma}{^x \Gamma}$$

$$\text{cut-rule} \quad \frac{A, \Gamma \quad \neg A, \Gamma}{\Gamma}$$

$$\text{structure rule} \quad \frac{\Gamma}{\Delta}, \text{ where } \Gamma^* \subseteq \Delta^*$$

Restriction. In the E1-rule, A is an atom and x any variable not having a free occurrence in A within the scope of an ε-symbol.

Notice that the β-rule, E1-rule, and cut rule all have two premisses and the remaining rules have only one. The structure rule is a very useful one since it permits us to add new members to a sequent, to rearrange the order of the members of a sequent, and to remove repetitions. Thus the sequent A, B, C follows from the sequent C, C, B by the structure rule. In each of the above rules the specified formula in the conclusion is called the *major formula*, and the specified formula(e) in the premiss(es) the *minor formula(e)*. Thus in the application of the β-rule by which $A \to B, C, D$ follows from $\neg A, C, D$ and B, C, D, the formula $A \to B$ is the major formula, and the formulae $\neg A$ and B are the minor formulae. An application of the structure rule has no major or minor formulae, and an application of the cut rule has no major formula. In any application of the cut rule the minor formula which was denoted above by A is called the *cut formula*.

EXERCISE

Let Γ, Δ_1, and Δ_2 be any three sequents. Suppose that Γ follows by one of the above rules of inference from Δ_1 (or from Δ_1 and Δ_2). Prove that if $\Delta_1^* \vdash_\varepsilon f$ (and $\Delta_2^* \vdash_\varepsilon f$), then $\Gamma^* \vdash_\varepsilon f$. (cf. Theorem II.12.)

2.1 Refutations

A *refutation* of the sequent Δ in the sequent calculus is any sequence of sequents such that the last member of the sequence is Δ and each member of the sequence either is an axiom or follows by some rule of inference from a preceding member (or from preceding members) of the sequence. A refutation in which the cut rule is not used is called a *normal refutation*. If there exists a refutation of Δ, we say Δ is *refutable* and denote this fact by writing $Ref(\Delta)$. Furthermore, if there exists a normal refutation of Δ, we write $norm\text{-}Ref(\Delta)$.

In order to show that the sequent calculus fits our general definition of a formal system, we define the notion of a deduction of A from X as follows: For any formula A (of $\mathcal{L}(\mathcal{V})$) and any set X of formulae (of $\mathcal{L}(\mathcal{V})$) the *deductions* of A from X in the sequent calculus (for $\mathcal{L}(\mathcal{V})$) are the refutations of sequents of the form $\neg A, \Delta$, where $\Delta^* \subseteq X$. Thus a *proof* of A is a refutation of the sequent $\neg A$.

By defining a refutation as a sequence of sequents we encounter a notational ambiguity, since the expression $\langle \Delta_1, \ldots, \Delta_n \rangle$ could mean the sequence whose members are the sequents Δ_1, and Δ_2, etc., or it could mean the sequence whose only member is the sequent $\Delta_1, \ldots, \Delta_n$. To avoid such ambiguity we shall denote a sequence of sequents by inserting a semi-colon between each member of the sequence. Thus the expression

$$\langle \Gamma_1; \ldots; \Gamma_m; \Delta, \Gamma \rangle$$

denotes a sequence of $m + 1$ sequents where for each $i = 1, \ldots, m$, Γ_i is the ith member, and the sequent Δ, Γ is the last member.

Having defined the notions of a refutation and a normal refutation, we can now give the precise statement of the major theorem of this chapter.

THE CUT ELIMINATION THEOREM. *For any ε-free sequent* Δ, *if Ref*(Δ), *then norm-Ref*(Δ).

In the next section we shall see why the notion of a normal refutation is so important.

3   The subformula property of normal refutations

The notion of an *immediate logical subformula* is defined by the following rules.

1. A is an immediate logical subformula of $\neg \neg A$.

2. For $i = 1, 2, \alpha_i$ is an immediate logical subformula of α, and β_i is an immediate logical subformula of β.

3. For any term t, $\gamma(t)$ is an immediate logical subformula of γ, and $\delta(t)$ is an immediate logical subformula of δ.

4. For any formulae A and B, A is an immediate logical subformula of B only if it is so by virtue of one of the above rules.

A formula A is said to be a *logical subformula* of B if and only if there exists a finite sequence of formulae whose first member is B and last member is A, such that every member of the sequence, except the first, is an immediate logical subformula of the preceding member. (Since the sequence may have only one member, then any formula is a logical subformula of itself.)

Now consider the rules of inference of the sequent calculus. Except in the case of the cut rule and the E1-rule, any formula which belongs to the premiss of a rule of inference is a logical subformula of some formula which belongs to the conclusion. In the case of the E1-rule every formula which belongs to the premiss is either the negation of an atom or a logical subformula of some formula which belongs to the conclusion. These facts yield the following *subformula property* of normal refutations:

If $\mathscr{R}$ is a normal refutation of Δ, then every formula in $\mathscr{R}$ (i.e., every formula which belongs to some sequent of $\mathscr{R}$) is either (i) the negation of an atom, or (ii) a logical subformula of some formula belonging to Δ.

Thus by proving the Cut Elimination Theorem, we can show that for every refutable ε-free Δ, there exists a refutation whose formulae are 'no more complicated' than those which belong to Δ.

3.1 The eliminability of the identity symbol

In order to demonstrate the usefulness of the subformula property we shall prove the following theorem concerning the eliminability of the identity-symbol from normal refutations.

A sequent is said to be *identity-free* if all its members are identity-free.

THEOREM V.1. *For any identity-free sequent Δ, if norm-Ref Δ, then there exists a normal refutation of Δ in which every sequent is identity-free.*

Proof. To prove the theorem it is sufficient to prove that there exists a normal refutation $\langle \Delta_1 ; \dots ; \Delta_n \rangle$ of Δ, where each Δ_i either is a C-axiom (see page 115) or follows by some rule other than the E1-rule. For in such a refutation, the identity symbol plays no essential role, and hence every occurrence of a quasi-formula of the form $s = t$ can be replaced by the formula f.

The proof hinges on the fact that if A is a subformula of some identity-free formula B, then its skeleton A^+ (p.66) is identity-free. Let $\langle \Gamma_1 ; \dots ; \Gamma_n \rangle$ be any normal refutation of Δ. By the subformula property of normal refutations, for any formula A in the Γ_i, if the skeleton of A is not identity free, then A must be the negation of an atom and hence of the form $\neg(s = t)$. Now for each $i = 1, \dots , n$ let Δ_i be obtained from Γ_i by replacing every member of Γ_i of the form $\neg(t = t)$ by f, and by removing every member of Γ_i of the form $\neg(s = t)$, where s and t are different. Then the sequence $\langle \Delta_1 ; \dots ; \Delta_n \rangle$ is the required refutation of Δ. For, if Γ_i is an axiom, then Δ_i is a C-axiom; if Γ_i follows from Γ_j and Γ_l by the E1-rule, then Δ_i follows by the structure rule from either Δ_j or Δ_l; and if Γ_i follows by any other rule of inference, then Δ_i follows by that same rule.

This theorem will be used later, in conjunction with the Cut Elimination Theorem, to give an alternative proof of the eliminability of the identity symbol from the predicate calculus.

4 The contradiction rule

In this section we shall prove the useful fact that if for some formula A, both A and $\neg A$ occur in Δ, then *norm-Ref*(Δ). The proof of this fact depends on the following notion of the index of a formula or term.

By the skeleton A^+ of any formula or term A we understand that expression which is obtained from A by replacing every subterm of A by a_1 (cf. p. 66). We now define the *index* of any formula or term to be the length of its skeleton, and we denote the index of A by $ind(A)$. Clearly the index function possesses the following properties:

K1. $ind(A) < ind(\neg A)$;
K2. If A is an immediate logical subformula of B, then $ind(A) < ind(B)$;
K3. if A' is obtained from A by replacing every occurrence of some term t by s, then $ind(A') = ind(A)$;
K4. if $\varepsilon y B$ is an ε-term of the form $[p]_t^x$ and p is subordinate to QxA, then $ind(\varepsilon y B) + 2 \leq ind(QxA)$.

Thus the index function, like the rank function in Chapter III, is a useful measure of the complexity of any term or formula.

THEOREM V.2 (The contradiction rule). *For any sequent Δ and any formula A, if A and $\neg A$ belong to Δ, then norm-Ref(Δ).*

Proof. The proof is by induction on the index of the formula $\neg A$. At each stage of the induction it is sufficient to prove either *norm-Ref*$(\neg A, A)$ or *norm-Ref*$(A, \neg A)$, since Δ follows from each of the sequents $A, \neg A$ and $\neg A, A$ by the structure rule. By our unifying classification of formulae (page 14) one of the following four cases must hold.
Case 1. A is an atom: In this case Δ is an axiom and therefore *norm-Ref*(Δ).
Case 2. A is of the form $\neg B$: Since $ind(\neg B) < ind(\neg A)$, we have *norm-Ref*$(B, \neg B)$ by the induction hypothesis. Hence *norm-Ref*$(\neg \neg B, \neg B)$ by the $\neg\neg$-rule.
Case 3. One of the two formulae, A and $\neg A$, is a conjunctive formula α and the other is a disjunctive formula β: By the duality principle (page 16) α_1 and β_1 are contradictory and α_2 and β_2 are contradictory. Hence by the induction hypothesis we have

$$norm\text{-}Ref(\beta_1, \alpha_1, \alpha_2) \quad \text{and}$$

$$norm\text{-}Ref(\beta_2, \alpha_1, \alpha_2).$$

Hence *norm-Ref*$(\beta, \alpha_1, \alpha_2)$ by the β-rule. The desired result, *norm-Ref*(α, β), now follows by applications of the structure rule, the α_1-rule, and the α_2-rule.
Case 4. One of the two formulae, A and $\neg A$, is a universal formula γ and the other is an existential formula δ: By the duality principle $\gamma(\varepsilon\delta)$ and $\delta(\varepsilon\delta)$ are contradictory. Hence by the induction hypothesis we have *norm-Ref*$(\gamma(\varepsilon\delta), \delta(\varepsilon\delta))$. The desired result, *norm-Ref*(δ, γ) now follows by applications of the γ-rule, structure rule, and δ-rule.

THEOREM V.3. *For any sequent Δ, if Δ^* is truth functionally invalid, then norm-Ref(Δ).*

M.L.—9

Proof. The proof is practically identical to the proof of Theorem II.13.

COROLLARY (The Tautology Theorem). *If A is a tautology, then norm-$Ref(\neg A)$.*

EXERCISE

Prove that if A is any axiom of the ε-calculus other than an E2-axiom, then *norm-$Ref(\neg A)$.*

4.1 The relationship between the sequent calculus and ε-calculus

THEOREM V.4. *For any sequent Δ and any formula A, $Ref(\neg A,\Delta)$ iff there exists a deduction of A from Δ^* in the ε-calculus in which no E2-axioms are used.*

Proof. (i) Let $\langle \Gamma_1 ; \ldots ; \Gamma_m \rangle$ be some refutation of $\neg A$, Δ. It is easy to prove by induction that for each $i = 1, \ldots, m$ there is a deduction of f from $\Gamma_i{}^*$ in the ε-calculus without the E2-axioms. (See the exercise at the end of § 2, page 116.) Hence $\{\neg A\} \cup \Delta^* \vdash_\varepsilon f$ (without E2), and therefore $\Delta^* \vdash_\varepsilon A$ (without E2) by the f-rule.

(ii) Conversely, assume that $\langle A_1, \ldots, A_n \rangle$ is a deduction of A from Δ^* in the ε-calculus (without E2). We shall prove by induction that for each $i = 1, \ldots, n$, $Ref(\neg A_i, \Delta)$. One of the following three cases must hold.

Case 1. A_i is an axiom (of the ε-calculus) other than an E2-axiom: Then $Ref(\neg A_i)$ by the exercise at the end of the last section. Hence $Ref(\neg A_i, \Delta)$ by the structure rule.

Case 2. A_i is a member of Δ^*: Then $Ref(\neg A_i, \Delta)$ by the contradiction rule.

Case 3. A_i follows from A_j and A_k by modus ponens: Then A_k is of the form $A_j \to A_i$. By the induction hypothesis we have

(1) $$Ref(\neg A_j, \Delta) \qquad \text{and}$$

(2) $$Ref(\neg(A_j \to A_i), \Delta).$$

The structure rule applied to (1) gives

(3) $$Ref(\neg A_j, \neg A_i, \Delta),$$

and by the contradiction rule, we have

(4) $$Ref(A_i, \neg A_i, \Delta).$$

The β-rule applied to (3) and (4) gives

(5) $$Ref(A_j \to A_i, \neg A_i, \Delta),$$

and the structure rule applied to (2) gives

(6) $$Ref(\neg(A_j \to A_i), \neg A_i, \Delta).$$

Finally, the cut rule applied to (5) and (6) gives

(7) $Ref(\neg A_i, \Delta)$.

THEOREM V.5. *Let* A *be any proper formula and* Δ *any sequent whose members are proper formulae. If norm-Ref($\neg A, \Delta$), then* $\Delta^* \vdash_{\varepsilon^*} A$.

Proof. Let $\langle \Gamma_1; \ldots; \Gamma_m \rangle$ be any normal refutation of $\neg A, \Delta$. Since every subformula of a proper formula is proper (and atoms are proper), then by the subformula property of normal refutations every member of Γ_i is proper, for each $i = 1, \ldots, n$. Hence, as in the proof of Theorem V.4(i), one can prove that $\Gamma_i^* \vdash_{\varepsilon^*} f$ for each $i = 1, \ldots, n$.

5 k, F_k-refutations

We now turn to the main problem, that of proving the eliminability of the cut rule. In order to prove this result by an inductive argument we need the following definition.

For any $k \geq 1$ and any finite collection F_k of formulae with index k, a k, F_k-*refutation* is a refutation $\langle \Gamma_1; \ldots; \Gamma_n \rangle$ such that a sequent Γ_i can follow by the cut rule from some Γ_j and Γ_l only if the cut formula A in that rule satisfies the following conditions: (i) $ind(A) \leq k$ and (ii) if $ind(A) = k$, then A is a member of F_k. If there exists a k, F_k-refutation of Δ, we write k, F_k-$Ref(\Delta)$. Obviously, if there exists a refutation of Δ, then there exists a $k \geq 1$ and a finite collection F_k of formula of index k such that k, F_k-$Ref(\Delta)$. Since every formula has index ≥ 1, then a 1, $\emptyset$-refutation is a normal refutation. Consequently, in order to prove the Cut Elimination Theorem it is sufficient to prove that if k, $\{A\} \cup F_k$-$Ref(\Delta)$, for some ε-free Δ, then k, F_k-$Ref(\Delta)$.

Since our definition of a k, F_k-refutation imposes a limit on the index of any formula which is the minor formula in an application of the cut rule, we have the following modified subformula property for k, F_k-refutations:

If $\mathscr{R}$ is a k, F_k-refutation of Δ, then for any formula A which belongs to some sequent of $\mathscr{R}$, either (i) A is a subformula of some member of Δ, or (ii) A is the negation of an atom, or (iii) $ind(A) \leq k + 1$.

5.1   The invertibility of the logical rules

Our next theorem states that the $\neg\neg$-rule, α-rules, β-rule, γ-rule, and δ-rule are, in a certain sense, invertible. The proof of this theorem depends on the fact that a non-atomic formula A can be the major formula in one and only one type of rule. For example, if A is a conjunctive formula, it can be the major formula only in the α-rules, and if A is disjunctive, only in the β-rule.

THEOREM V.6. *For any sequent* Δ, *and any formula* A, *conjunctive formula* α, *disjunctive formula* β, *universal formula* γ, *and existential formula* δ:

(i) $k, F_k\text{-}Ref(\neg\neg A, \Delta)$ iff $k, F_k\text{-}Ref(A, \Delta)$;

(ii) $k, F_k\text{-}Ref(\alpha, \Delta)$ iff $k, F_k\text{-}Ref(\alpha_1, \alpha_2, \Delta)$;

(iii) $k, F_k\text{-}Ref(\beta, \Delta)$ iff $k, F_k\text{-}Ref(\beta_1, \Delta)$ and $k, F_k\text{-}Ref(\beta_2, \Delta)$;

(iv) $k, F_k\text{-}Ref(\gamma, \Delta)$ iff there exists terms $t_1, \ldots, t_n$ such that
$$k, F_k\text{-}Ref(\gamma(t_1), \ldots, \gamma(t_n), \Delta);$$

(v) $k, F_k\text{-}Ref(\delta, \Delta)$ iff $k, F_k\text{-}Ref(\delta(\varepsilon\delta), \Delta)$.

Proof. We shall prove only part (iv), since the other four parts can be proved in a similar fashion.

(iv) First assume $k, F_k\text{-}Ref(\gamma(t_1), \ldots, \gamma(t_n), \Delta)$. Then by repeated applications of the γ-rule and the structure rule we obtain a k, F_k-refutation of γ, Δ.

Conversely, let $\langle \Gamma_1; \ldots; \Gamma_m \rangle$ be a k, F_k refutation of γ, Δ. For notational simplicity suppose that γ is the formula $\forall x A$. (The proof is identical for the case where γ is of the form $\neg \exists x A$.) Let $t_1, \ldots, t_n$ be all those terms t such that, for some $j = 1, \ldots, m$, Γ_j is of the form $A(t), \Lambda$. (If there are no such t, let t_1 be a_1.) For any sequent Λ, let Λ^0 denote the sequent obtained from Λ by removing every occurrence of $\forall x A$. For each $i = 1, \ldots, m$ let Γ_i' be the sequent $\Gamma_i^0, A(t_1), \ldots, A(t_n)$. We now claim that the sequence of sequents

$$\langle \Gamma_1'; \ldots; \Gamma_m'; A(t_1), \ldots, A(t_n), \Delta \rangle$$

is a k, F_k-refutation of $A(t_1), \ldots, A(t_n), \Delta$. Since Γ_m' is the sequent $\Delta^0, A(t_1), \ldots, A(t_n)$, then the final sequent $A(t_1), \ldots, A(t_n), \Delta$ follows from Γ_m' by the structure rule. The justifications for the Γ_i' are as follows. If Γ_i is an axiom, then clearly Γ_i' is also an axiom since $\forall x A$ is neither an atom nor the negation of an atom. Suppose Γ_i follows from Γ_j (and Γ_l) by some rule of inference R. One of the following cases must hold.

Case 1. The formula $\forall x A$ is neither a major nor minor formula in this application of the rule R: Then Γ_i' follows from Γ_j' (and Γ_l') by the same rule R. (This includes the case where R is the structure rule.)

Case 2. The formula $\forall x A$ is the major formula in this application of the rule R: Then R *must be* the γ-rule, Γ_i is of the form $\forall x A, \Lambda$, and Γ_j of the form $A(t), \Lambda$. Hence t is one of the terms $t_1, \ldots, t_n$, and therefore Γ_i' follows from Γ_j' by the structure rule.

Case 3. The formula $\forall x A$ is a minor formula in this application of the rule R: Then Γ_i' follows from Γ_j' (or Γ_l') by the structure rule.

EXERCISE

Show how the required refutations for the other four parts of Theorem V.6 can be constructed.

The technique employed in proving Theorem V.6 can be used to prove the following theorem.

THEOREM V.7. *For any atom* A, *if* k, F_k-$Ref(\neg A,\Delta)$ *and* k, F_k-$Ref(A,\Delta)$, *then* k, F_k-$Ref(\Delta)$.

Proof. (Note: we cannot apply the cut rule to prove this theorem since the index of A may be greater than k.)

Case 1. A is the formula f: Let $\langle \Gamma_1 ; \ldots ; \Gamma_m \rangle$ be any k, F_k-refutation of $\neg f, \Delta$. For each $i = 1, \ldots, m$, let Γ_i^0 be the sequent obtained from Γ_i by removing each occurrence of $\neg f$. Then $\langle \Gamma_1^0 ; \ldots ; \Gamma_m^0 ; \Delta \rangle$ is a k, F_k-refutation of Δ.

Case 2. A is not the formula f: Let $\langle \Gamma_1 ; \ldots ; \Gamma_m \rangle$ be any k, F_k-refutation of A,Δ and $\langle \Lambda_1 ; \ldots ; \Lambda_n \rangle$ any k, F_k-refutation of $\neg A, \Delta$. For each $i = 1, \ldots, m$, let Γ_i^0 be the sequent obtained from Γ_i by removing each occurrence of A and let Γ_i' be the sequent Γ_i^0, Δ. Then the sequence

$$\langle \Lambda_1 ; \ldots ; \Lambda_n ; \Gamma_1' ; \ldots ; \Gamma_m' ; \Delta \rangle$$

is a k, F_k-refutation of Δ. The proof of this assertion is identical to the proof of Theorem V.6, except for the following two changes: (1) Case 2 in the proof of Theorem V.6 does not apply since no rule of inference has an atom as its major formula, and (2) if Γ_i is an axiom by virtue of the fact that both A and $\neg A$ occur in it, then Γ_i' follows from Λ_n (i.e., $\neg A,\Delta$) by the structure rule.

In order to prove the Cut Elimination Theorem we would like to improve on Theorem V.6(v) by showing that if k, F_k-$Ref(\delta,\Delta)$, then for any term t, k, F_k-$Ref(\delta(t),\Delta)$. The following theorem shows that this is indeed the case provided that certain conditions are satisfied.

THEOREM V.8. *Let δ be any existential formula and Δ any ε-free sequent. Suppose k, F_k-$Ref(\delta,\Delta)$, where $ind(\delta) \geqslant k$ and the length of δ is at least as great as the length of any member of F_k. Then for any term t,*

$$k, \ F_k\text{-}Ref(\delta(t),\Delta).$$

Proof. Since k, F_k-$Ref(\delta,\Delta)$, then by Theorem V.6(v) k, F_k-$Ref(\delta(\varepsilon\delta),\Delta)$. Recall that if δ is $\exists y B$, then $\delta(\varepsilon\delta)$ is $B(\varepsilon y B)$, and if δ is $\neg \forall y C$, then $\delta(\varepsilon\delta)$ is $\neg C(\varepsilon y \neg C)$. Thus the term denoted by $\varepsilon\delta$ has the same index and the same length as δ. Let $\delta(\varepsilon\delta)$ be the formula $B(\varepsilon y B)$ and let $\langle \Gamma_1 ; \ldots ; \Gamma_m \rangle$ be the k, F_k-refutation of $B(\varepsilon y B),\Delta$. Notice that by our assumption on the length of δ the term $\varepsilon y B$ does not occur in any member of F_k. For each $i = 1, \ldots, m$, let Γ_i^0 be the sequent that we obtain from Γ_i by replacing each occurrence of $\varepsilon y B$ (in the members of Γ_i) by the term t. Let Γ_i' be the sequent $\Gamma_i^0, B(t)$. We shall prove that the sequence $\langle \Gamma_1' ; \ldots ; \Gamma_m' ; B(t),\Delta \rangle$ is a k, F_k-refutation of $B(t), \Delta$, i.e., of $\delta(t), \Delta$.

First of all, since Δ is ε-free, then Γ_m' is $B(t), \Delta, B(t)$, so that $B(t), \Delta$ follows from Γ_m' by the structure rule. If Γ_i is an axiom, then Γ_i' is also an axiom, and if Γ_i follows from Γ_j (and Γ_l) by some rule of inference other than the γ-rule or δ-rule, then clearly Γ_i' follows from Γ_j' (and Γ_l') by the same rule.

We now consider the case where Γ_i follows from Γ_j by the γ-rule or δ-rule. First of all, suppose that Γ_j is of the form $B(\varepsilon y B)$, Λ. Then Γ_j' is $B(t)$, Λ^0, $B(t)$, and Γ_i' follows from Γ_j' by the structure rule. On the other hand, suppose Γ_j is of the form $A(s)$, Λ, where $A(s)$ is not $B(\varepsilon y B)$. Assume Γ_i is of the form $\forall x A$, Λ. (The other possibilities can be handled similarly.) As in the proof of Theorem III.6, we want to rule out the possibility that $\varepsilon y B$ is of the form $[p]_t^x$ where p is a quasi ε-term subordinate to $\forall x A$. Suppose that this is the case. Then the index of $\forall x A$ is at least $k + 2$ (since the index of $\varepsilon y B$ is at least k). Consequently, by the subformula property of k, F_k-refutations, $\forall x A$ must be a subformula of some member of Δ. But this is impossible since every member of Δ is ε-free, and an improper formula, such as $\forall x A$, cannot be a subformula of an ε-free formula. Hence Γ_i' is of the form $\forall x A^0$, Λ^0, $B(t)$, and Γ_j' is of the form $A^0(s^0)$, Λ^0, $B(t)$; and Γ_i' follows from Γ_j' by the γ-rule. This completes the proof. (Why is this refutation still a k, F_k-refutation?)

COROLLARY. *For any ε-free Δ, any existential formula δ, and any term t, if norm-Ref(δ,Δ), then norm-Ref($\delta(t)$,Δ).*

EXERCISE

Using the cut rule give a simple proof of the following: for any sequent Δ, any existential formula δ, and any term t, if Ref(δ,Δ), then Ref($\delta(t)$,Δ). Using this result prove that Theorem V.8 still holds if we remove the condition that $ind(\delta) \geq k$.

6   The Cut Elimination Theorem

The next theorem is the last subsidiary result needed in the proof of the Cut Elimination Theorem.

THEOREM V.9 (The cut property). *For each $i = 1, \ldots, n$, let A_i and B_i be contradictory formulae with index $\leq k$. If k, F_k-Ref($A_1, \ldots, A_n$, Δ) and k, F_k-Ref(B_i,Δ) for each $i = 1, \ldots, n$, then k, F_k-Ref(Δ).*

Proof. The proof is by induction on n.
Case 1. $n = 1$: Assume k, F_k-Ref(A,Δ) and k, F_k-Ref(B,Δ), where A and B are contradictory and each has index $\leq k$. Then Δ follows from A, Δ and B, Δ by the cut rule, where the cut formula has index $< k$ (since its negation has index $\leq k$), and hence k, F_k-Ref(Δ).
Case 2. $n \geqslant 2$: Assume k, F_k-Ref($A_1, \ldots, A_n$,Δ) and k, F_k-Ref(B_i,Δ) for each $i = 1, \ldots, n$. By the structure rule, from k, F_k-Ref(B_1,Δ) we infer

k, F_k-$Ref(B_1, A_2, \ldots, A_n, \Delta)$. Case 1 then gives k, F_k-$Ref(A_2, \ldots, A_n, \Delta)$. The desired result now follows by the induction hypothesis.

THEOREM V.10 (The Cut Elimination Theorem). *For any ε-free sequent* Δ, *if Ref(Δ), then norm-Ref(Δ).*

Proof. By using an obvious inductive argument, we can reduce this theorem to the following lemma.

LEMMA. *Suppose* k, $\{A\} \cup F_k$-$Ref(\Delta)$, *where* Δ *is ε-free and A is any formula with index k whose length is at least as great as the length of any member of* F_k. *Then* k, F_k-$Ref(\Delta)$.

Proof. Let $\langle \Gamma_1; \ldots; \Gamma_m \rangle$ be a k, $\{A\} \cup F_k$-refutation of Δ. For each $i = 1, \ldots, m$, let Γ_i' be Γ_i, A, and let Γ_i'' be Γ_i, $\neg A$. Then the sequence

$$\langle \Gamma_1'; \ldots, \Gamma_m'; A, \Delta \rangle$$

is a k, F_k-refutation of A, Δ, and the sequence

$$\langle \Gamma_1''; \ldots; \Gamma_m''; \neg A, \Delta \rangle$$

is a k, F_k-refutation of $\neg A$, Δ. For, if Γ_i follows from Γ_j and Γ_l by an application of the cut rule where A and $\neg A$ are the minor formulae, then Γ_i' and Γ_i'' follow by the structure rule. Otherwise, Γ_i' and Γ_i'' have the same justification that Γ_i has in the original refutation. Hence we have

(1) k, F_k-$Ref(A, \Delta)$,   and

(2) k, F_k-$Ref(\neg A, \Delta)$.

As in the proof of Theorem V.2, one of the following cases must hold.
Case 1. A is an atom: Then by Theorem V.7, (1) and (2) yield k, F_k-Ref(Δ).
Case 2. A is of the form $\neg B$: By Theorem V.6(i), (2) yields k, F_k-$Ref(B, \Delta)$. Since $ind(\neg B) = k$, then by (1) and the cut property we have k, F_k-$Ref(\Delta)$.
Case 3. One of the two formulae, A and $\neg A$, *is a conjunctive formula* α *and the other is a disjunctive formula* β: Thus we have

(3) k, F_k-$Ref(\alpha, \Delta)$,   and

(4) k, F_k-$Ref(\beta, \Delta)$.

By Theorem V.6, (3) and (4) yield

(5) k, F_k-$Ref(\alpha_1, \alpha_2, \Delta)$,

(6) k, F_k-$Ref(\beta_1, \Delta)$,   and

(7) k, F_k-$Ref(\beta_2, \Delta)$.

Since α_1 and β_1 are contradictory and α_2 and β_2 are contradictory, and the index of each of these formulae is less than k, then the cut property yields k, F_k-$Ref(\Delta)$.

Case 4. One of the two formulae, A and $\neg A$, is a universal formulae γ and the other is an existential formula δ: Thus we have

(8) $k, F_k\text{-}Ref(\gamma,\Delta)$,   and

(9) $k, F_k\text{-}Ref(\delta,\Delta)$.

By (8) and Theorem V.6(iv) there exist terms $t_1, \ldots, t_n$ such that

(10) $k, F_k\text{-}Ref(\gamma(t_1), \ldots, \gamma(t_n),\Delta)$.

Since Δ is a ε-free and since δ is either A or $\neg A$ we can apply Theorem V.8 to (9) and get for each $i = 1, \ldots, n$

(11) $k, F_k\text{-}Ref(\delta(t_i),\Delta)$.

The desired result, $k, F_k\text{-}Ref(\Delta)$ now follows from (10) and (11) by the cut property.

This completes the proof of the lemma and hence also of Theorem V.10.

The following counterexample shows that the Cut Elimination Theorem no longer holds if the condition that Δ is ε-free is removed.

Let P be any 1-place predicate symbol. Since the sequent $\exists x Px$, $\neg P\varepsilon x Px$ follows by the δ-rule from the axiom $P\varepsilon x Px$, $\neg P\varepsilon x Px$, we have at once *norm-Ref*$(\exists x Px, \neg P\varepsilon x Px)$. Hence, by the exercise at the end of the last section, there exists a refutation of Pt, $\neg P\varepsilon x Px$ for any term t. However, since Pt and $P\varepsilon x Px$ are both atoms, it is easy to see that there exists no *normal* refutation of the sequent Pt, $\neg P\varepsilon x Px$ when t is any term other than $\varepsilon x Px$. Consequently, if we let Δ be the sequent Pa, $\neg P\varepsilon x Px$ the Cut Elimination Theorem does not hold. (For possible ways of modifying the axioms and rules of inference of the sequent calculus so that the Cut Elimination Theorem holds for arbitrary Δ, see Maehara [1955], [1957] and Curry [1963], page 342.)

6.1 Applications

In this section we show how the Cut Elimination Theorem can be used to provide new proofs of some of our earlier results. In each case the main function of this theorem is to prove the eliminability of certain symbols or certain types of formulae.

THEOREM V.11 (The Second ε-Theorem—weaker form). *For any ε-free X and A, if $X \vdash_\varepsilon A$ (without* E2), *then $X \vdash_{PC} A$.*

Proof. Since $X \vdash_\varepsilon A$ (without E2), then there exist formulae $B_1, \ldots, B_n$ such that $B_1, \ldots, B_n \vdash_\varepsilon A$ (without E2). Let Δ be the sequent $B_1, \ldots, B_n$. Then $Ref(\neg A,\Delta)$ by Theorem V.4. Hence *norm-Ref*$(\neg A,\Delta)$ by the Cut

Elimination Theorem and therefore $X \vdash_{\varepsilon^*} A$ by Theorem V.5. This implies $X \vdash_{PC} A$ by Theorem III.3.

Notice that the Cut Elimination Theorem has been used in this proof to show that any improper formulae can be eliminated from the original deduction of A from X. (cf. Theorem V.5.)

THEOREM V.12. *For any ε-free X and A, $X \vdash_{PC} A$ iff there exists a sequent Δ, such that $\Delta^* \subseteq X$ and norm-Ref($\neg A, \Delta$).*

Proof. Assume $X \vdash_{PC} A$. Then $X \vdash_{\varepsilon^*} A$ by Theorem III.2. Hence *norm-Ref*($\neg A, \Delta$) as in the proof of Theorem V.11. Conversely, if *norm-Ref*($\neg A, \Delta$), where $\Delta^* \subseteq X$, then $X \vdash_{PC} A$ as in the proof of Theorem V.11.

Using this essential link between deductions in the predicate calculus and normal refutations, we can now give new proofs of the First ε-Theorem and of the eliminability of the identity symbol (cf. Theorem III.10 and Theorem III.15).

THEOREM V.13. *Let X be any set of ε-free $\forall$-prenex formulae and B any $\exists$-prenex formula. If $X \vdash_{PC} B$, then $A_1, \ldots, A_m \vdash_{EC} B_1 \vee \ldots \vee B_n$ where the A_i are certain substitution instances of the matrices of members of Y and the B_j are certain substitution instances of the matrix of B.*

Proof. By Theorem V.12, there exists a sequent Δ such that $\Delta^* \subseteq X$ and *norm-Ref*($\neg B, \Delta$). By repeated applications of the invertibility of the γ-rule (Theorem V.6(iv)) we get *norm-Ref*($\neg B_1, \ldots, \neg B_m, A_1, \ldots, A_n$), where the B_j and A_i are of the required form. By applying the α-rules and the structure rule, this yields

$$norm\text{-}Ref(\neg(B_1 \vee \ldots \vee B_n), A_1, \ldots, A_n).$$

Now since each of the A_i and B_j are quantifier-free, then by the subformula property of normal refutations every member of the refutation is elementary (since any stray ε-terms can be replaced by a). Consequently, as in the proof of Theorem V.5 we get

$$A_1, \ldots, A_m \vdash_{EC} B_1 \vee \ldots \vee B_n.$$

THEOREM V.14. *For any identity-free (ε-free) X and A, if $X \vdash_{PC} A$, then there exists a deduction of A from X in the predicate calculus without identity.*

Proof. Since $X \vdash_{PC} A$, then by Theorem V.12, *norm-Ref*($\neg A, \Delta$) for some $\Delta^* \subseteq X$. This implies by Theorem V.1 that there exists a normal refutation of $\neg A$, Δ in which every sequent is identity-free. Returning now to the predicate calculus by means of Theorem V.12, we obtain an identity-free deduction of A from X.

BIBLIOGRAPHY

ACKERMANN, W.
   1924. 'Begründung des "tertium non datur" mittels der Hilbertschen Theorie der Widerspruchsfreiheit', *Mathematische Annalen*, **93**, 1–36.
   1937–8. 'Mengentheoretische Begründung der Logik', *Mathematische Annalen*, **115**, 1–22.
   1940. 'Zur Widerspruchsfreiheit der Zahlertheorie', *Mathematische Annalen*, **117**, 162–194.

ASSER, G.
   1957. 'Theorie der logischen Auswahlfunktionen', *Zeitschrift für mathematische Logik und Grundlagen der Mathematik, 3.* 30–68.

BERNAYS, P.
   1927. 'Zusatz zu Hilberts Vortrag über "Die Grundlagen der Mathematik" ', *Abhandlungen aus dem mathematischen Seminar der Hamburgischen Universität*, **6**, 89–92. (cf. von Heijenoort [1967], pp. 485–489.)
   1935. 'Hilberts Untersuchungen über die Grundlagen der Arithmetik', in Hilbert [1935], pp. 196–216.
   1958. *Axiomatic Set Theory*, Amsterdam. (With a historical introduction by A. A. Fraenkel.)

BETH, E. W.
   1959. *The Foundations of Mathematics*, Amsterdam.

BOURBAKI, N.
   1954. *Éléments de Mathématique, Livre I (Théorie des Ensembles), Chap. I et II*, Paris.

CARNAP, R.
   1961. 'On the use of Hilbert's ε-operator in scientific theories', *Essays on the Foundations of Mathematics dedicated to A. A. Fraenkel on his Seventieth Anniversary*, Jerusalem, pp. 156–164.

CHURCH, A.
   1956. *Introduction to Mathematical Logic, Volume I*, Princeton.

COHEN, P. J.
   1963. *The independence of the axiom of choice* (mimeographed, Stanford University).
   1966. *Set Theory and the Continuum Hypothesis*, New York.

CURRY, H. B.
   1963. *Foundations of Mathematical Logic*, New York.

130

DENTON, J., and DREBEN, B.
1969. 'The Herbrand theorem and the consistency of arithmetic'. Forthcoming.

DREBEN, B., ANDREWS, P., and AANDERAA, S.
1963. 'False lemmas in Herbrand', *Bulletin of the American Mathematical Society*, **69**, 699–706.

FRAENKEL, A. A., and BAR-HILLEL, Y.
1958. *Foundations of Set Theory*, Amsterdam.

FRAYNE, T., MOREL, A. C., and SCOTT, D. S.
1962. 'Reduced direct products', *Fundamenta Mathematicae*, **51**, 195–228.

GENTZEN, G.
1934–5. 'Untersuchungen über das logischen Schließen', *Mathematische Zeitschrift*, **39**, 176–210, 405–431.
1936. 'Die Widerspruchsfreiheit der reinen Zahlentheorie', *Mathematische Annalen*, **112**, 493–565.

GÖDEL, K.
1931. 'Über formal unentscheidbare Sätze der Principia Mathematica und verwandter Systeme I', *Monatshefte für Mathematik und Physik*, **38**, 173–198.
1940. *The Consistency of the Axiom of Choice and of the Generalized Continuum Hypothesis with the Axioms of Set Theory*, Princeton.

HAILPERIN, T.
1957. 'A theory of restricted quantification, I and II', *Journal of Symbolic Logic*, **22**, 19–35 and 113–129.

van HEIJENOORT, J.
1967. (editor) *From Frege to Gödel, A Source Book in Mathematical Logic*, 1879–1931. Cambridge, Mass.

HENKIN, L.
1949. 'The completeness of the first-order functional calculus', *Journal of Symbolic Logic*, **14**, 159–166.

HERBRAND, J.
1930. 'Recherches sur la théorie de la démonstration', *Travaux de la Société des Sciences et des Lettres de Varsovie*, Classe III sciences mathématiques et physiques, no. 33, pp. 33–160.
1931. Sur la non-contradiction de l'arithmétique, *Journal für die reine und angewandte Mathematik*, **166**, 1–8. (cf. van Heijenoort [1967], pp. 618–628.)

HERMES, H.
1965. *Eine Termlogik mit Auswahloperator*, Berlin-Heidelberg-New York.

HILBERT, D.
1923. 'Die logischen Grundlagen der Mathematik', *Mathematische Annalen*, **88**, 151–165.

1926. 'Über das Unendliche', *Mathematische Annalen*, **95**, 161–190. (cf. van Heijenoort [1967], pp. 367–392.)

1928. 'Die Grundlagen der Mathematik', *Abhandlungen aus dem Mathematischen Seminar der Hamburgischen Universität*, **6**, 65–85. (cf. van Heijenoort [1967], pp. 464–479.)

1935. *Gesammelte Abhandlungen,* vol. 3, Berlin.

HILBERT, D., and BERNAYS, P.
1934. *Grundlagen der Mathematik*, vol. 1, Berlin. Reprinted Ann Arbor, Mich., 1944.
1939. *Grundlagen der Mathematik*, vol. 2, Berlin. Reprinted Ann Arbor, Mich., 1944.

KLEENE, S. C.
1952. *Introduction to Metamathematics*, Amsterdam, Groningen, New York.

KNEEBONE, G. T.
1963. *Mathematical Logic and the Foundations of Mathematics*, London.

KREISEL, G.
1964. 'Hilbert's Programme', *Philosophy of Mathematics*, (edited by P. Benacerraf and H. Putnam), Oxford, pp. 157–180.

LEISENRING, A. C.
1968. 'An abstract property of formalized languages which contain Hilbert's ε-symbol', *Zeitschrift für mathematische Logik und Grundlagen der Mathematik*, **14**, 81–92.

LÉVY, A.
1961. 'Comparing the axioms of local and universal choice', *Essays on the Foundations of Mathematics dedicated to A. A. Fraenkel on his Seventieth Anniversary*, Jerusalem, pp. 83–90.

LYNDON, R. C.
1966. *Notes on Logic*, Princeton.

MAEHARA, S.
1955. 'The predicate calculus with ε-symbol,' *Journal of the Mathematical Society of Japan*, **7**, 323–344.
1957. 'Equality axioms on Hilbert's ε-symbol', *Journal of the Faculty of Science, University of Tokyo*, Sect. 1, **7**, 419–435.

MENDELSON, E.
1964. *Introduction to Mathematical Logic*, Princeton.

MONTAGUE, R., and HENKIN, L.
1956. 'On the definition of formal deduction', *Journal of Symbolic Logic*, **21**, 129–136.

von NEUMANN, J.
1927. 'Zur Hilbertschen Beweistheorie', *Mathematische Zeitschrift*, **26**, 1–46.

132　　　　　　　　BIBLIOGRAPHY

QUINE, W. V.
　　1950. 'On natural deduction', *Journal of Symbolic Logic*, **15**, 93–102.

RASIOWA, H.
　　1956. 'On the ε-theorems', *Fundamenta Mathematicae*, **43**, 156–165.

ROBINSON, A.
　　1963. *Introduction to Model Theory and the Metamathematics of Algebra*, Amsterdam.

SCHOENFIELD, J. R.
　　1954. 'A relative consistency proof', *Journal of Symbolic Logic*, **19**, 21–28.
　　1967. *Mathematical Logic*, Reading, Mass.

SMULLYAN, R.
　　1965. 'Analytic natural deduction', *Journal of Symbolic Logic*, **30**, 123–139.
　　1966a. 'Trees and nest structures', *Ibid.*, **31**, 303–321.
　　1966b. 'Finite nest structures and propositional logic', *Ibid.*, **31**, 322–324.

TAIT, W. W.
　　1965. 'The substitution method', *Journal of Symbolic Logic*, **30**, 175–192.

TARSKI, A.
　　1930. 'Fundamentale Begriffe der Methodologie der deduktiven Wissenschaften, I', *Monatshefte für Mathematik und Physik*, **37**, 361–404.

WANG, H.
　　1955. 'On denumerable bases of formal systems', *Mathematical Interpretations of Formal Systems*, Amsterdam, pp. 57–84.
　　1963. 'The predicate calculus', *A Survey of Mathematical Logic*, Amsterdam, pp. 307–321.

INDEX OF SYMBOLS

INDEX OF RULES OF
INFERENCE

(The numbers refer to the pages on which the rules are introduced.)

GENERAL INDEX